PRECISION AGRICULTURE FOR GRAIN PRODUCTION SYSTEMS

Brett Whelan and James Taylor

PUBLISHING

National Library of Australia Cataloguing-in-Publication entry

Whelan, Brett.

Precision agriculture for grain production systems/by Brett Whelan and James Taylor.

9780643107472 (pbk.)
9780643107489 (epdf)
9780643107496 (epub)

Includes bibliographical references and index.

Precision farming – Study and teaching – Australia.
Agriculture – Management – Australia.
Farmers – Education – Australia.

Taylor, James Arnold.

630.710994

Published by

CSIRO PUBLISHING
36 Gardiner Road, Clayton VIC 3168
Private Bag 10, Clayton South VIC 3169
Australia

Telephone: [+613] 9545 8555
Local call: 1300 788 000 (Australia only)
Fax: +61 3 9662 7555
Email: csiropublishing@csiro.au
Web site: www.publishing.csiro.au

Front cover: Variability in production everywhere you look (Brett Whelan)

Set in Adobe Minion Pro 11/13.5 and Adobe Helvetica Neue LT
Edited by Adrienne de Kretser, Righting Writing
Cover design by Jenny Cowan
Text design by James Kelly
Typeset by Desktop Concepts Pty Ltd, Melbourne
Index by Russell Brooks
Printed by Ingram Lightning Source

Feb26_RP_ILS

Contents

Preface

The expanding human population, projected resource limitations and tougher environmental regulations are exerting an increasing pressure on crop production systems. This pressure is driving investigation and investment in new technologies and techniques that aim to increase total production, while optimising production efficiency and addressing the growing environmental concerns of society.

Precision Agriculture (PA) offers suitable channels for such investments, and has reached a level of development where it can provide useful technologies and techniques that support the targeting of these goals.

The authors, with the financial support of the Grains Research and Development Corporation, have been extensively involved in the research, development and application of PA in grain crop production. The collaboration has produced this book, which aims to provide an understanding of the principles that underpin the major technologies and techniques being used in PA, and uses production examples to explain their applications and value in grain crop management. The concepts, and many of the tools described here, will also have relevance to most other agricultural industries.

While PA in the grains industry has been built on a long history of innovators and pioneers, exciting challenges lie ahead for PA practitioners in areas such as:

- fine-scale, real-time, cost-effective estimation of soil profile/crop nutrients;
- fine-scale, real-time, cost-effective estimation of soil profile moisture content;
- localised weather predictions;
- efficient, integrated crop quality monitors;
- spatial yield prediction/simulation models;
- combining crop reflectance sensors with an independent biomass sensor;
- better understanding of the agronomic impact of fine-scale resource variability and interactions;
- autonomous weeding;
- targeting PA for increased water-use efficiency and improved farm C and N emission management;

- improving PA GIS capabilities;
- improved integration of multiple data layers for real-time decision-making in nutrient/irrigation applications;
- product tracking and production information traceability;
- secondary and tertiary education.

As the research efforts continue and the promising PA technologies and techniques are adopted and adapted to suit local requirements and conditions, it is easy to see that the PA philosophy will become a crucial component in sustainably (commercially and environmentally) managing all inputs, natural retentions and emissions across inherently variable agricultural enterprises.

It is this inclusion into fundamental cropping management that will herald the true success of PA and reward the efforts of the inquisitive and innovative crop producers, consultants and research colleagues who have given generously of their time, expertise and resources to contribute to the development of PA.

Finally, we would especially like to acknowledge the farmers who have been directly involved in our research, and the significant contribution and support from Professor Alex McBratney and the many research students and staff that have been part of the PA Team at the University of Sydney.

Brett Whelan and James Taylor

1

Introduction to Precision Agriculture

Precision Agriculture (PA) is a now a term used throughout agricultural systems worldwide, but what is meant by Precision Agriculture? This introductory chapter provides a background to the principal philosophy and goals of a PA management strategy, the evolution of PA and some of the steps required to adopt PA in grain cropping systems. It provides a stepping-stone to subsequent chapters that will investigate the theories, technologies and methodologies behind the adoption of PA within grain production systems.

Defining PA

Many definitions of PA exist and many people have different ideas of what PA should encompass. Here, we have selected two definitions to illustrate the concept of PA in general and specifically its application to broadacre cropping industries. The first definition comes from the US House of Representatives (US House of Representatives 1997).

Precision Agriculture

An integrated information- and production-based farming system that is designed to increase long-term, site-specific and whole-farm production efficiency, productivity and profitability while minimising unintended impacts on wildlife and the environment.

The key to this definition is that it identifies PA as a 'whole-farm' management strategy, not just a strategy for individual fields. It uses information technology to improve production management and minimise environmental impact. This definition refers to the farming system that in modern agriculture may include the supply chain from the farm gate to the consumer; it also distinguishes between agriculture and agronomy. While the PA philosophy has been expounded primarily in cropping industries, it is important to remember that PA can relate to any agricultural production system. These may involve animal industries, horticulture, viticulture, fisheries and forestry; in many cases PA techniques are being implemented without being identified as such. For example, in a dairy enterprise the tailoring of feed allocations to individual cows may be based on yield and the stage of their lactation. The focus in this book is the implementation of PA in grain production.

The second definition narrows the PA philosophy of timely management of variation down to its implementation in cropping systems.

Site-specific crop management (SSCM)

A form of PA whereby decisions on resource application and agronomic practices are improved to better match soil and crop requirements as they vary in the field.

This definition encompasses the idea that PA is an evolving management strategy. The focus here is on decision-making with regard to resource use and not necessarily to the adoption of information technology on farm, although many new technologies will aid improved decision-making. The decisions can be in regard to changes across a field at a certain time in the season, or to changes through a season or seasons. The inference is that better decision-making will provide a wide range of benefits – economic, environmental and social – that may or may not be known or measurable at present (Figure 1.1).

From a grain production perspective, this definition provides a goal regardless of a grower's current adoption of PA or proposed entry level into PA. Expanding this goal, SSCM can be considered as applying information at the site-specific level, with grower knowledge, to achieve the objectives of:

- optimising production efficiency;
- optimising quality;
- minimising environmental impact;
- minimising risk.

Objectives of SSCM

The success of an SSCM strategy will depend on how each or all of the above objectives are met.

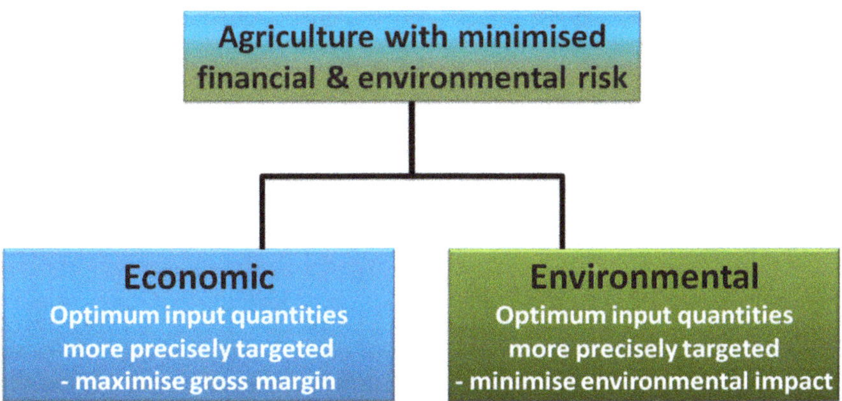

Figure 1.1: The economic–environmental foundation for an SSCM to minimise operational and societal risk.

Optimising production efficiency

In general, the aim of SSCM is to optimise returns across a field. Rarely do fields have uniform yield potential, so identifying any variation in yield potential may offer the possibility of optimising yield production by varying management within fields. The initial emphasis should be on optimising the agronomic response to the manageable input that has the greatest impact on production and costs. In the absence of any clear environmental benefits, this will be achieved by differentially applying inputs up to the point that the marginal return (MR) = marginal cost (MC) at each site or in each region of the field that has been identified as having a different yield potential.

Optimising quality

In general, production efficiency is measured in terms of a yield (quantity) response, mainly because yield and biomass sensors are the most reliable and commonplace sensors. In the past few years the first attempts to commercialise grain quality sensors have been made and on-the-go grain protein/oil sensors are now commercially available. The ability to site-specifically collect grain quality data will allow growers to consider production efficiency from the perspective of yield or quality, or a yield–quality interaction. Many inputs will impact on quality as well as quantity. In production systems where quality premiums exist this may alter the amount of input required to optimise profitability and agronomic response.

In some product markets, where strong quality premiums/penalties are applied, a uniform approach to quality properties may be optimal. The quality of some agricultural commodities is greatly increased by decreasing the variability in production, e.g. malting barley. If quality premiums more than offset yield loss then growers may prefer to vary inputs to achieve uniform production quality (minimise variability in quality), rather than optimise yield productivity.

Minimising environmental impact

From an environmental point of view, more precise treatments may offer the prospect of reducing the environmental risk associated with uniform/blanket field treatments and provide the ability to work with the natural diversity in cropping systems.

If management decisions are tailoring inputs to meet production needs, then by default there must be a decrease in the net loss of any applied input to the environment. This is not to say that there is no actual or potential environmental damage associated with the production system; however, the risk of environmental damage is reduced.

By recording the amount and location of any input application, producers have physical evidence to contest any claims regarding negligent management. They can also provide information on 'considerate' practices, to gain market advantage. A by-product of improved information collection and flow is a general improvement in producers' understanding of the production system and the potential implications of different management options.

Apart from avoiding litigation or chasing product segmentation into markets, there is currently little regulatory incentive for growers to capture and use information on the environmental footprint of their production in Australia. Other countries, particularly within the European Union (EU), are financially encouraging producers to collect and use this information by linking environmental issues to subsidy payments. Should such eco-service payments be introduced in Australia, the value of PA could be further enhanced.

Minimising risk

Risk management is a common practice for most farmers and can be considered from two points of view – income and environment. In a production system, farmers often practise risk management by erring on the side of extra inputs while the unit cost of a particular input is deemed 'low'. Thus, a farmer may apply an extra agrochemical spray, add extra fertiliser, buy more machinery or hire extra labour to ensure that the produce is grown/harvested/sold on time, thereby guaranteeing a return.

Generally, minimising income risk is seen as more important than minimising environmental risk. SSCM attempts to offer a risk management solution that may allow both positions to be considered. This improved management strategy depends on a better understanding of the environment–crop interaction and a more detailed use of emerging and existing information technologies (e.g. short- and long-term weather predictions and agro-economic modelling).

The more that is known about a production system, the faster a producer can adapt to changes. Many inputs will impact on quality as well as quantity. In production systems where quality premiums exist, this may alter the amount of input required to optimise profitability and agronomic response. More information on the level of production quality may also permit a grower to better target products within the supply chain and minimise or better manage off-farm risks.

Implementation of SSCM

The SSCM cycle can be described with five key nodes (Figure 1.2). A brief introduction is provided here; further details are covered in later chapters. It is important to remember that SSCM is a continuous management strategy. Initially, some form of monitoring and data analysis are needed to make a decision. However, it is equally important to continue to monitor and analyse the effect of that decision, and feed that information into subsequent management decisions.

Geo-referencing

The ability to geo-reference data, i.e. link it to a specific location on the Earth's surface, is the truly enabling technology of SSCM in its present form. Global navigation satellite systems (GNSS), of which the global positioning system (GPS) is the most widely used at present, are now common on many farms. Receivers, and the systems in which they are used, range in accuracy from ±10–20 m to ±2–3 cm and in price from A$200 to A$40 000. Applications include crop and soil monitoring, yield mapping and vehicle autosteer systems. The technology continues to improve and the price of receivers continues to decrease.

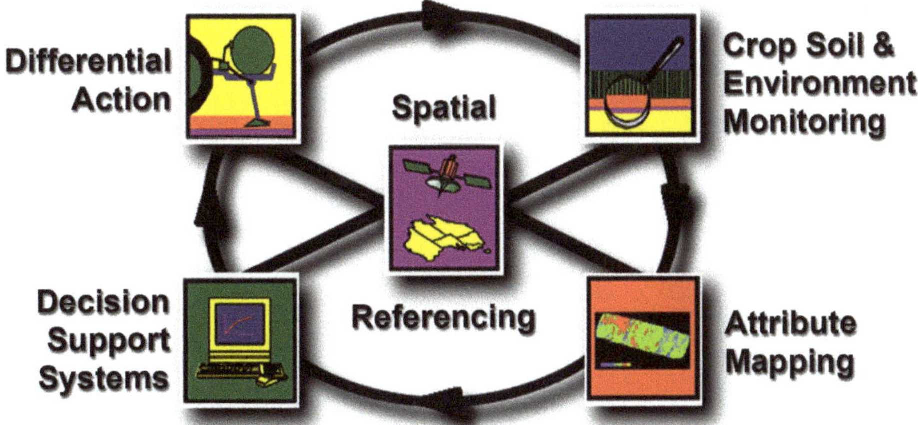

Figure 1.2: The SSCM cycle, showing that spatial referencing is the enabling technology that drives the other parts of the cycle.

The ability to link a location to an action or data allows producers to map and visually display farm operations. This provides insights into production variability as well as inefficiencies in crop production and farm operations. Recently, the more accurate GNSS have become more common on-farm as growers embrace vehicle guidance and autosteer technologies. These permit vehicles to travel along repeatable paths with minimum overlap, reducing driver fatigue and permitting greater timeliness in operations.

Crop, soil and climate monitoring

Many sensors and monitors already exist for *in situ* and on-the-go measurement for a variety of crop, soil and climatic variables. These include yield sensors, biomass and crop response sensors (aerial and space-borne multi- and hyper-spectral cameras), radio or mobile phone networked weather stations, apparent soil electrical conductivity (EC_a) sensors and gamma-radiometric soil sensors, to name a few. The majority of SSCM research in Australia is directed at identifying how to use the output from these sensors to improve production.

The other challenge for SSCM is to adapt *in situ* sensors and develop new on-the-go sensors. The commercial potential of these sensors means that private industry will be keen to focus on the engineering aspects of research and development, but research bodies have an important role to play in the development of the science behind the sensors to ensure they provide valid results, relevant to Australian production systems.

Attribute mapping

Crop, soil and climate sensors often produce large, spatially intensive data sets. The observations are usually irregularly spaced and need to be 'cleaned' and mapped onto a continuous surface to permit analysis. For several decades, geo-statisticians have been researching ways of describing and presenting spatial data that accurately represents the raw data. Software specifically designed for mapping and displaying agricultural data from different sources on a common platform is improving annually.

Decision support systems

Decision support systems (DSS) use agronomic and environmental data, combined with information on possible management techniques, to determine the optimum management strategy for production. Most commercial cropping DSS are based on 'average' crop response across a whole field and are therefore not designed for SSCM.

Similarly, the majority of engineering companies supplying SSCM technologies do not produce detailed DSS to support the use of their equipment in a commercial production system. This is substantially due to the highly site-specific nature of the factors affecting crop response, and the onus to fill this gap falls on external sources.

Initially, it may be sufficient to try to adapt existing cropping DSS to site-specific applications of inputs. In the long run, a DSS that is able to site-specifically model plant–environment interactions in terms of yield and quality will be needed. It must be flexible enough to incorporate a variety of sensor-gathered data, able to accept feedback from other parts of the SSCM cycle and conform to relevant ISO standards.

In the meantime, farmers and advisors are gathering data on variation, conducting within-field input response experiments if necessary, and using this information to augment current decision-making processes.

Differential action

A management operation where more than one level of a treatment or activity is undertaken is referred to as a differential action. The use of differential management requires information on where, when and by how much actions should be varied.

The actual differential application of inputs using variable-rate technology (VRT) is not a complex engineering issue. Based on this and the commercial potential, VRT for fertiliser application was probably the best-developed technology used in the SSCM cycle in the early 2000s. Development of new and refined methods for differential action remains a project for many research and commercial entities around the globe. Like GNSS receivers, VRT is becoming more user-friendly, more cost-effective and more common, especially in broadacre agriculture. The biggest barrier to adoption remains the lack of local information on where, when and by how much inputs should be varied.

A brief development history of SSCM

SSCM is not a particularly new concept in agriculture – essays on the topic date from the early 18th century. What is new is the scale at which SSCM can be implemented, using modern technologies.

Prior to the industrial revolution, agriculture was generally conducted on small fields. Farmers often had a detailed knowledge of their production system without actually quantifying the variability. The movement towards mechanical agriculture, and the profit margin squeeze, resulted in the latter half of the 20th century being dominated by large-scale uniform 'average' agricultural practices. The advance of technology in the late 20th and early 21st centuries has allowed agriculture to move back towards site-specific agriculture while retaining the economies of scale associated with large operations.

The impetus for the current concept of PA in cropping systems emerged in the late 1980s, when grid-based sampling of soil chemical properties could be matched to newly developed VRT for fertilisers. Using a compass and dead-reckoning principles, fertilisers were applied at rates designed to complement changes in soil fertility maps. Crop yield monitoring technologies were still in the research phase. Around 1990, the NAVSTAR GPS became the first GNSS available in a limited capacity for civilian use, and the opportunity for rapid and accurate vehicle location and navigation sparked a flurry of activity.

Electronic controllers for variable-rate application (VRA) were built to handle this new positioning information and crop yield monitors began to hit the commercial market. By 1993, the GPS was fully operational and a number of crop yield monitoring systems allowed the fine-scale monitoring and mapping of yield variation within fields. The linking of yield variability data at this scale with maps of soil nutrient changes across a field marked the true beginning of PA in broadacre cropping.

As yield monitoring systems were improved, it became evident that methods other than grid sampling for corroborative information would need to be developed. In many instances, grid sampling at the intensity required to correctly characterise variability in soil and crop parameters proved cost-prohibitive and, by the late 1990s, a 'management class' approach to sample-site selection had become a real option for management. This approach uses already-held information on variability to subdivide whole fields into areas that can be classed as having a similar crop/yield response. Samples are then extracted to characterise the difference between the classes. This technique helps adjust for current limitations in data quality and resolution, while trying to minimise costs and maximise the extraction of relevant information from sampling surveys.

The management class system should be considered an interim approach in the full development of SSCM. New systems and instruments for measuring or inferring soil and crop parameters on a more continuous basis are being developed, using both proximal (ground-based) and remote (aerial and satellite) platforms. Examples are instruments that measure EC_a using electromagnetic induction, crop reflectance imaging systems and crop quality sensors. As the ability to continuously measure variability at a fine scale increases, the capital cost of VRT decreases and the environmental value is factored in, SSCM will move to a truly site-specific management approach (Figure 1.3).

The success, and the potential for further success, of SSCM in the grains industry has prompted other farming industries, particularly viticultural and horticultural crops, to adopt PA. Since the late 1990s, research into non-grain crops has increased. More emphasis is being placed on the environmental auditing capabilities of PA technologies and their potential use in improving product traceability. Advances in GNSS technology since 1999 have opened the door for vehicle guidance, autosteering and controlled-traffic farming (CTF). CTF has provided sustainability benefits (e.g. minimisation of soil compaction), economic benefits (by minimising input overlap and improving timeliness of operations) and

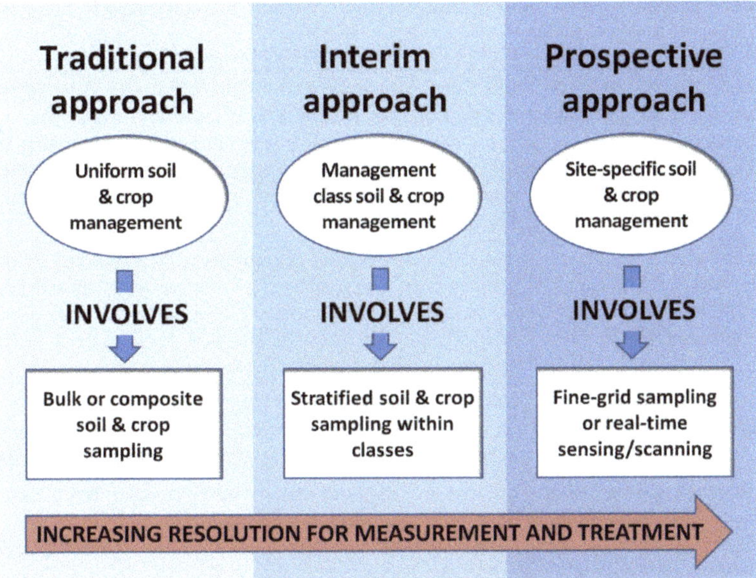

Figure 1.3: The evolution of SSCM from a uniform to a totally site-specific approach.

social benefits (e.g. reducing driver fatigue). As a result, this form of PA technology showed swift adoption rates in the first decade of the 21st century.

General adoption path

The application of the PA philosophy will have some costs, so the decision to move down the PA path should be made only after any current uniform farm management has been optimised. Larger gains may well be possible if some of the uniform agronomic practices are corrected before a producer adopts PA to fine-tune production management (Table 1.1).

Key points

- Precision Agriculture (PA) is essentially a philosophy for making improved management decisions across a farm at more places and more times in the season than has traditionally been the practice in agriculture.
- PA can potentially benefit many aspects of a farming operation: production (increased input-use efficiency), environmental (reduced chemical and physical footprint) and supply chain (improved quality and value-adding options).
- Site-specific crop management (SSCM) involves the application of PA to crop management: decisions on resource application and agronomic practices are improved to better match soil and crop requirements as they change in a field.

Table 1.1: Generalised stages for the progressive adoption of PA in grain production

	Objective	How PA tools and techniques may be used
1.	Ensure uniform-rate crop management is optimised and improve farming efficiency	Use GPS navigation and location recording for crop scouting, soil sampling and simple field experimentation (e.g. varieties or input rates). Use vehicle guidance and autosteering to increase efficiency (e.g. reduced overlap), establish controlled traffic or raised bed systems, sow into inter-rows, perform shielded spraying etc.
2.	Measure where and how much crop production varies across fields, the farm and seasons	Use GPS-linked yield quantity/quality monitors, other on-ground crop and soil sensors and remotely sensed data to develop maps of variability. Compare maps over seasons and crops to identify patterns and size of production variability across fields and whole farms.
3.	Determine the major causes of the variability	Use maps and images to target field investigation into the likely causes of variability in production. Soil and crop tissue testing in areas highlighted by the maps will help pinpoint causes.
4.	Optimise the use of inputs to maximise gross margin and minimise environmental footprint	May involve delineating different management classes at a field or farm scale. Use the information already gathered to formulate required actions. May involve adapting standard agronomic practices, in-field experimentation or applying the results of scenario simulation models to test alternative management options. These include ameliorating for problems (e.g. pH) or preparing variable-rate input application maps if required. The use of in-season crop reflectance sensors may provide useful information for nitrogen management strategies. Use yield monitors and financial analysis to check results and underpin further adaptive management.
5.	Improve grain quality control and product marketing	Use GPS with grain quality monitors to meet quality requirements and achieve price premiums. Electronic information tagging of field operations and grain loads provides data to support marketing and quality control/assurance.

- SSCM can be applied to management at the individual field or whole-farm scale.
- SSCM is an evolving management strategy. It is driving a shift from uniform management of inputs towards management by defined production classes and on to continuously variable management.
- Management by production classes deals with identifying and treating areas with different production potentials. Management classes are a common way of compressing many data layers and implementing SSCM.
- Identification of management classes within the field and the application of SSCM techniques and technologies should be validated by relevant crop response measurements (e.g. yield or quality).
- Production variation may be due to environmental factors or management factors. It may occur in crop quantity (yield) and quality, and can be managed in both.
- On-farm experimentation is an integral part of PA and is used for increasing knowledge of a production system.

- There are opportunities to manage both temporal (between seasons) and spatial (within-field) variations in crop production.
- PA adoption starts when a farmer wants to apply a new level of knowledge to a production process. It is not necessarily synonymous with buying a GPS unit and/or a yield monitor.
- There are various entry points into PA for farmers. Not all of them involve a large initial capital outlay.
- Differences in soil and climate variability between cropping regions in Australia mean that the most suitable/viable PA techniques and technologies may differ between regions. Producers should always consider the use of equipment and techniques in light of their specific issue.

2

Global Navigation Satellite Systems and Precision Agriculture

Satellite-based navigation systems are truly the enabling technology of PA. They provide a relatively simple and robust technique for identifying any location on the Earth's surface or, in the case of aircraft, relative to the surface. This permits agricultural and environmental operations to be geo-referenced and spatially analysed. A wide range of satellite-based navigation and geo-location tools are available to suit different agronomic situations from point crop/soil sampling to autonomous vehicle guidance.

Introduction to GNSS

Global navigation satellite system (GNSS) is the standard industry term for satellite-based navigation systems that provide geo-spatial positioning on or near the Earth's surface. While the term GPS has entered the modern vernacular to describe such systems, GPS is actually the moniker for the US NAVSTAR global positioning system. Since the mid 1990s the GPS has been the only fully operational worldwide GNSS. Consequently, the majority of receivers use the information from the NAVSTAR GPS satellites, so this is the system of most interest to agriculture at present.

In addition to NAVSTAR GPS, there is the Russian-operated Global'naya Navigatsionnaya Sputnikovaya Sistema (GLONASS). It fell into disrepair in the 1990s but has undergone a full restoration, reaching worldwide coverage at the end of 2011. Both GPS and GLONASS were launched towards the end of the Cold War, primarily for military use. Only later were they adapted for civilian use. They are termed GNSS-1 systems.

The widespread adoption of GPS receivers over the past 10–15 years by non-military users has prompted the development of new-generation GNSS, termed GNSS-2 systems, which have a greater focus on civilian applications. The most anticipated of the GNSS-2 systems is the EU's GALILEO, projected to be in intermediate operation in 2015 and fully operational by 2020.

The main difference between GNSS-1 and GNSS-2 systems is the quality of the radio signals. The GNSS-1 systems require a second augmentation signal (a differential correction signal) to produce high-quality positioning results for civilian users. The GNSS-2 systems will have increased accuracy without a differential signal and will carry information on more bands. The NAVSTAR GPS system is being overhauled to improve its signal quality to match GNSS-2 system specifications at present.

As well as global systems, there are several regional satellite-based navigation systems in operation (e.g. Beidou China) or under development (e.g. QZSS, Japan and IRNSS, India). These regional systems tend to use a constellation of geo-stationary satellites rather than a constellation of orbiting satellites like GPS and GLONASS, so they have applications only in specific regions and often for specific purposes.

All GNSS have three common components:

- the space segment, comprising the satellites;
- the control segment, which monitors and maintains the satellites;
- the user segment, which consists of receivers that use signals from the satellites to calculate location on or near the Earth's surface.

As end-users, civilians have a choice only of which receiver to use within the user segment. No civilian users have the ability to influence either the space or control segments. The detail in this chapter will be based on the GPS, but all GNSS operate in conceptually the same manner.

Transmissions from GPS satellites

Each GPS navigation satellite continuously broadcasts information on two L-band radio frequencies (L1 and L2).

- The L1 band carries two ranging codes, the coarse acquisition (C/A) code and precision (P) code. The C/A code is the signal used to measure distances in most basic civilian operations. The P code can only be decoded by military GPS units.
- The L1 band also contains a navigation message that supplies precise information on the individual satellite's orbit and clock correction (the

ephemeris data) so that the satellite's position in space relative to the centre of the Earth can be calculated by a receiver. There is also coarser information on the paths of all the satellites, the general transmission errors expected and general information on the health of the whole satellite system (the *almanac* data).
- The L2-band carries only the P code.

Both the C/A and P signals are digital codes that are unique for each satellite. They are repeatedly sent from the satellites with a time of transmission stamp. Receivers have a store of the unique codes generated by the satellites. When a receiver intercepts the digital code from a satellite, the satellite can be identified and its location in space calculated from the associated navigation message.

Calculating a receiver location

A GPS receiver uses the information gathered from the satellites to provide location information at a given time. A GPS receiver can do this because the satellite signals being monitored are travelling at the speed of light, which is a constant 299 792.5 km/s. When the duration of time taken for a signal to travel is determined, the distance between the satellite and the receiver can be calculated as follows:

$$\text{Travel time (s)} \times \text{Speed of travel (ms}^{-1}) = \text{Distance travelled (m)} \quad \text{(Eq. 2.1)}$$

When this is done for a number of satellites and the information on the location of each satellite in space relative to the Earth is determined, the data can be used to geo-locate a receiver, using trilateration.

To truly geo-locate itself, a receiver needs to determine its latitude, longitude and elevation. This is known as a position in three dimensions (3D). To do this, a receiver needs to be able to 'see' and access data from at least four satellites. Most GPS receivers will not give a reading unless four satellites are being simultaneously tracked (Figure 2.1).

In 'code-phase' GPS operation, the C/A ranging code being carried on the L1-band radio wave is used to determine the signal travel time from each satellite. When a code arrives at the receiver, the reception time is compared with the stated time of transmission and the difference is the time taken for the signal to travel between the two. At present, code-phase operation is restricted to the L1 band unless the user has a military decoder that can read the P code (which is on both radio waves). Calculating travel times using information from the code on only a single radio wave restricts the degree of accuracy that can be achieved. This is due to the relatively long cycle time of the code transmission (1/1000th of a second) and errors introduced by location and time-variable delays on signal transmission by

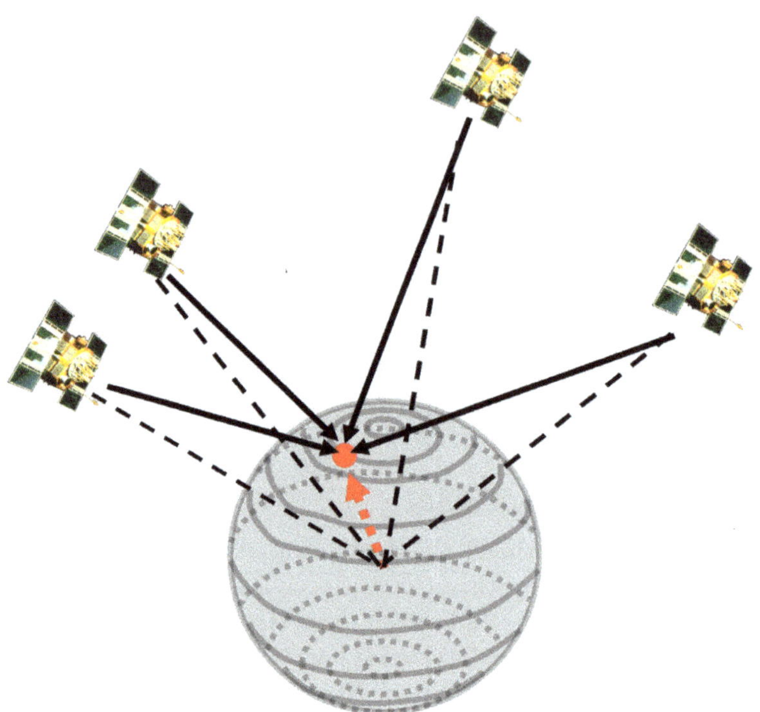

Figure 2.1: Schematic of GPS location determination. Positions of at least four satellites, relative to Earth centre, are calculated from broadcast ephemerides. Distances from the receiver to the satellites are calculated and used, in conjunction with satellite positions, to converge to a single point on the surface of the Earth.

the Earth's atmosphere (see 'GPS errors'). Using information from codes on two frequencies can eliminate this error.

'Carrier-phase' operation is a way of improving the accuracy of time, and therefore position, measurement without a military decoder. For this, the actual radio wave or 'carrier' is used to improve the time measurement process. If two carriers are monitored the transmission delay errors can be removed. Using the carrier in the timing process improves time measurement because the carrier is cycling 1000 times faster than the code it carries, so the smallest unit of time in the measurement process is 1000 times smaller. Monitoring the difference in time measurements calculated from two carriers of different frequency enables the atmospheric delay to be determined.

Only advanced GPS units are able to operate in carrier-phase mode.

Turning on a GPS receiver

When a GPS receiver is turned on, what occurs depends on when and where it was last operated. All receivers need to know the time and to acquire the almanac data for the system along with the ephemeris data from each satellite that will be used for geo-location. The almanac is transmitted every 12.5 minutes by all satellites and is considered valid for several months. The ephemeris data is unique for each satellite, is transmitted every 30 seconds and considered valid for 30 minutes.

A new receiver, a receiver that has been unused for many months or one that has been moved more than 300 km will have no valid information about the correct time, its location, which satellites should be 'visible' or where they are in the sky. It will need to perform a cold start.

Cold start

- Perform a full C/A code and frequency search to find a satellite.
- Once one satellite is located, time is synchronised and the almanac will be downloaded.
- Three satellites will be located and an approximate receiver location will then be computed.
- The ephemeris data will be downloaded from at least four satellites to enable full 3D positioning.
- The whole process should take less than 15 minutes.

A receiver that has a valid last position stored, along with valid almanac information so that the identity and expected locations of satellites is known, will perform a warm start.

Warm start

- Three satellites will be located and the approximate receiver location computed.
- The ephemeris data will be downloaded from at least four satellites to enable full 3D positioning.
- The whole process should take less than a minute.

A receiver that has a current ephemeris, almanac, receiver location and time but that has lost satellite signal/s due to line-of-sight interference issues will perform a re-acquisition.

Re-acquisition or hot start

- Lock will be re-established on the necessary satellite/s signal to enable full 3D positioning.
- The whole process should take less than four seconds once the interference is removed.

GPS errors

The accuracy of a receiver is influenced by many parameters, not just its mode of operation. It is important to know what may be affecting signal quality, particularly for high-accuracy applications. Apart from the mode of operation used in the timing process (C/A code versus P code versus carrier-phase) the error in position calculated by a receiver may be affected by a number of error sources. By far the biggest error is loss of line-of-sight between the receiver and the satellites. For this reason, receivers work less well indoors or under thick foliage. This is particularly important to remember when using GPS in fields with trees that may disrupt the signal. Apart from a loss of line-of-sight, the main errors in positioning accuracy are satellite geometry, satellite errors, atmospheric errors, multipath errors, receiver errors and continental drift errors.

Satellite geometry

Apart from errors in determining the distance between the satellites and receiver, the accuracy of geo-location is also a function of the geometry of the satellites used for geo-location. The optimum geometry is for one satellite to be directly overhead and the other three spread out evenly. As satellites orbit the Earth, their geometry relative to a receiver varies and the dilution of position (DOP) errors will vary; this is the main cause of daily variation in the accuracy of geo-location. Receivers with upwards of 12 satellite tracking channels help minimise this effect.

Satellite errors

These can be errors in the time settings of the onboard atomic clocks (timing error) or an error in the transmitted data about the location of the satellite in space (ephemeris error).

Atmospheric errors

To reach a GPS receiver, the satellite signal needs to pass through the Earth's atmosphere, in particular the ionosphere and troposphere that degrade or slow the signal speed.

Multipath errors

These are errors caused when the GPS antenna receives signals that have been reflected from a secondary source, such as nearby sheds or silos. This lengthens the travel time and thus creates error in the distance determination. Multipath errors can be minimised by not placing a fixed base station antenna near buildings and by operating mobile systems away from large structures.

Receiver errors

The ability of the GPS receiver and associated software to cope with thermal and electronic noise from external sources, such as motors, may affect how accurately the receiver can geo-locate itself. The receiver clock, which is not as accurate as the atomic clocks in the satellites, may also add error.

Continental drift errors

Australia drifts approximately 7 cm in a north-easterly direction every year. Differential GPS (DGPS) is required to overcome this problem and ensure repeatability from one season to the next.

Differential correction

External sources of error combine to create a total error in the position calculation due to factors other than the receiver and its mode of operation. It is possible to estimate this external error by recording the GPS signal at a fixed surveyed location (known as a base station) then comparing the GPS position calculated by the base station to its known location. This calculated difference can be used to improve the position calculations of nearby mobile receivers. When such a correction is used in position determination, the operational mode is known as 'augmented' or 'differential' GPS.

For static or point positioning operations, the correction can be recorded separately and the position corrected later (post-processing). More commonly in agriculture, with mostly mobile positioning requirements, the correction can be transmitted to a mobile receiver by a radio or internet link and applied in real time.

Types of GPS receivers

Stand-alone GPS receivers

This term literally means a receiver that operates with no external correction. However, it is commonly used to refer to standard position service (SPS) receivers, which operate using only the basic C/A code on the L1 band from the navigation satellites. These are the cheapest GPS receivers as they have no facility for incorporating a correction signal or complex circuitry to use the P code or carrier-phase mode of operation. SPS receivers have the lowest positioning accuracy of all the GPS receivers on the market. Most SPS receivers contain filtering algorithms designed to smooth the signal noise when the GPS is moving. This makes SPS receivers more accurate when moving and suitable for wide-swathing, low-resolution applications such as yield monitoring.

Real-time differential GPS (DGPS) receivers

These require two antennas (often not obvious in the hardware): one to collect the signal from the GPS satellites and determine a position and another to receive a correction signal, containing a correction factor, to improve the accuracy of the position. The correction signal can come from a variety of sources that provide different levels of accuracy. The main options are a local base station, a coastguard beacon or a wide area differential GNSS.

Local base station

Any user can establish a local base station using a second GPS receiver and a pair of radios to transmit and receive the correction signal. This is known as operating in Real-Time Kinematic (RTK) mode. Using a single, fixed base station for correction assumes that all errors applying at the reference station apply equally to the mobile receiver. Therefore, the effects of ionospheric delay can be compensated for to a degree. The tropospheric errors are only minimally corrected in single base station DGPS. As the distance between the two receivers increases, the receivers begin to observe different satellite information errors and receive the satellite signals via different travel paths through the atmosphere.

Figure 2.2 shows the basic set-up of a real-time DGPS using a local base station where:

- the mobile receiver (A) and the fixed position base station (B) receive positioning data from the navigation satellites (C);
- the base station (B) continually compares its surveyed position with that calculated using the data from the satellites (C);
- a correction algorithm (differential) is computed to adjust the incoming data and the differential is relayed by radio frequency to the mobile unit (A);
- the mobile unit is able to calculate its position with greater accuracy using data from the satellites (C) and the differential supplied by the base station (B).

Generally, single local base stations are limited to a range of less than 30 km by the strength of the radio signals used and the possibility of terrain interference. Some alternative approaches to signal delivery have been used in an effort to make the signal more accessible. However, issues with correction signal degradation need to be understood when using these wider area signals.

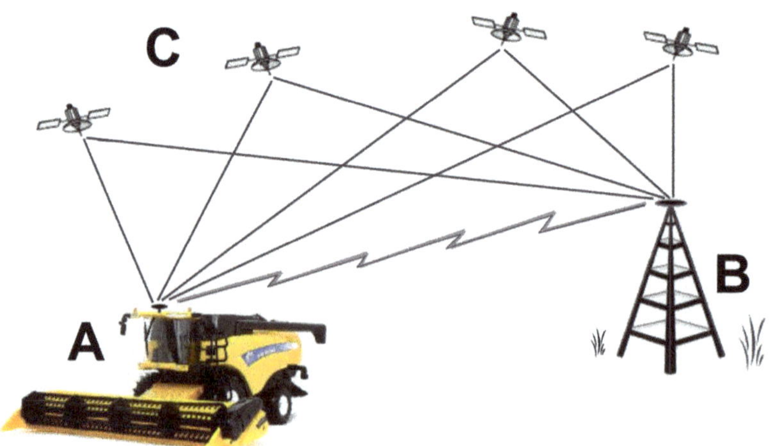

Figure 2.2: Schematic of a real-time DGPS using a local base station.

Coastguard beacon

It is possible for some farming areas in Victoria, South Australia and Queensland to receive a free correction signal from a maritime navigation beacon system. There are plans to install more maritime beacons along the east coast of Australia in the coming years, but their effective range will depend on user proximity to the coast, intervening terrain and the signal strength of each beacon.

Wide area differential GNSS (WADGNSS)

Limitations with radio signal and loss of precision by moving away from the base station can be overcome by using two or more base stations. The corrections from numerous base stations can be combined into a correction algorithm that is optimised for any user located within the base station network.

With these systems, the differential corrections computed at various ground stations are analysed and combined to obtain an improved estimate of errors at a user's location. The calculated correction improves with additional base stations.

As shown in Figure 2.3, a network of fixed position receivers (D) communicates with the GPS satellites. Each calculates a correction algorithm for its location, which is then passed to a master station (E). The master station computes a system-wide correction algorithm that varies the correction adjustment for a GPS position based on a user's location within the network. This correction algorithm is relayed to a general communications satellite (F) that increases the broadcast range to remote users (G). The correction transmission is supplied in a standard format defined by the Radio Technical Commission for Maritime Services (RTCM-104).

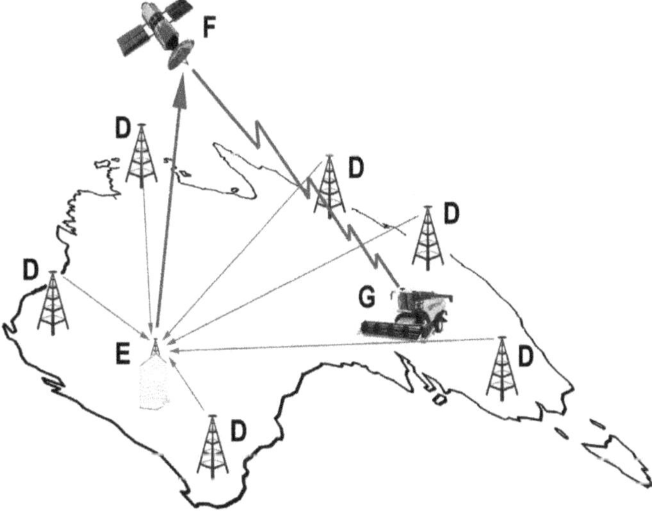

Figure 2.3: Schematic of a real-time wide area differential system (WADGPS).

This process is often referred to as creating a virtual base station for any user wherever the receiver may travel within the network. This form of correction is provided commercially by OmniSTAR and John Deere, with sub-metre and decimetre precision services available by subscription across Australia.

WADGNSS services available in Australia

John Deere Ltd

SF1 World Solution: This is the base-level differential correction signal for John Deere receivers. SF1 delivers ±25 cm pass-to-pass precision (95% of the time).

SF2: A subscription-based upgrade from the basic correction, the SF2 signal delivers ±10 cm pass-to-pass precision (95% of the time).

OmniSTAR/Trimble

Virtual base station (VBS): VBS is a worldwide subscription service capable of providing sub-metre (< ±1 m) positioning accuracy. Data from the available base stations within the network are processed together using a weighted, least squares approach. The corrections are then adjusted for the user's specific position within the network.

OmniSTAR-HP: A subscription service that uses dual-frequency carrier-phase measurements to provide the positioning corrections. By using dual-frequency GPS receivers the ionospheric delay can be measured at the base and user locations, allowing the error to be removed. The OmniSTAR-HP service provides positioning to within ±10 cm (95% of the time).

OmniSTAR-XP: OmniSTAR-XP takes the HP solution and makes it available anywhere in the world irrespective of the user's proximity to a reference station. This subscription service combines the OmniSTAR-HP dual-frequency carrier-phase solution with clock and orbit corrections for each GPS satellite. This provides positioning to within ±50 cm (95% of the time) anywhere in the world that an OmniSTAR L-band broadcast can be received.

Continually operating reference stations (CORS) network

CORS are essentially local base stations that have been surveyed into their true position and can provide a continuous local correction via wireless communication (e.g. internet, 3G/4G mobile phone). When these stations are networked, a regional correction solution can be established. Depending on how far apart the base stations are, the horizontal accuracy of the correction can range from more than 1 m (local WADGNSS CORS) to about 2 cm (NRTK CORS). To establish a NRTK CORS regional correction network, the stations cannot be more than 50–70 km apart.

All state governments in Australia have begun establishing CORS networks for use by subscription. Coverage began in the capital cities and is slowly including

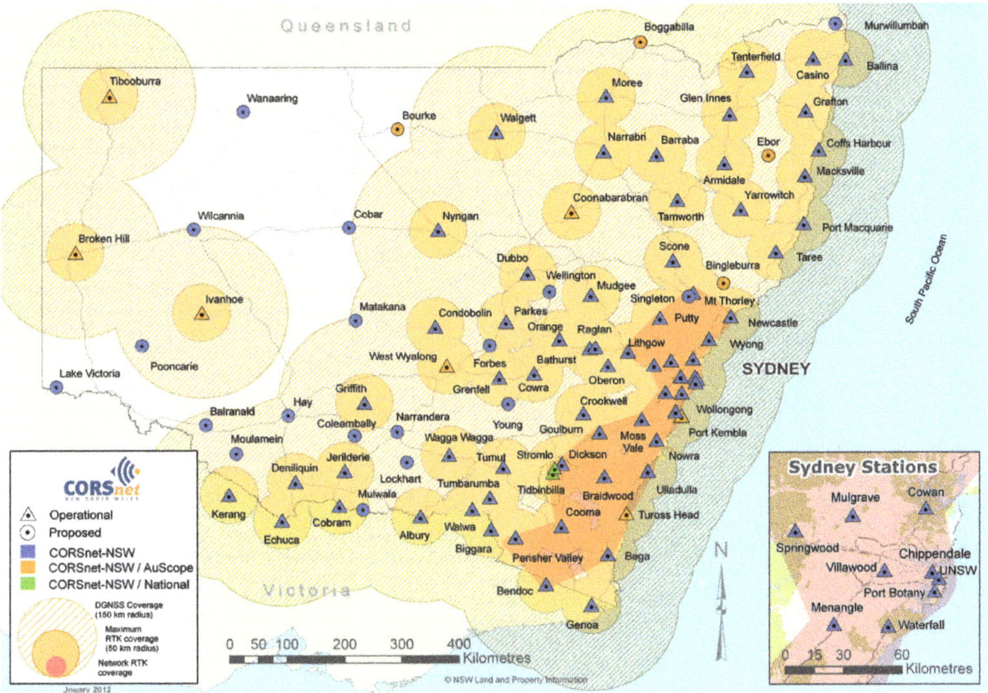

Figure 2.4: Operational and proposed configuration of the NSW CORSnet network (2012).

agricultural regions (Figure 2.4). The most widespread network at present is GPSnet (Victoria), which is being used by the agricultural community. A number of private or industry CORS networks are also being established around Australia for agricultural use.

DGPS operational modes

DGPS receivers tend to be more expensive than the stand-alone GPS receivers, as they require extra components to accept the correction signal and update the position. These receivers can operate in C/A code or carrier-phase or in a combination of both. Depending on the receiver mode of operation and the quality of the correction applied, the accuracy can range from ±2 mm to ±2 m.

Carrier-phase DGPS receivers use the carrier-phase mode of operation to determine an initial position with greater accuracy. Carrier-phase systems may be either single-frequency (accessing only the L1 band signals) or dual-frequency (accessing both L1 and L2 band signals). Dual-frequency receivers have the advantages of faster acquisition time and the ability to better correct initially for atmospheric signal travel delay. Some are also capable of accessing GLONASS as well as GPS satellites, if required. Any correction information used at this level must come from other carrier-phase GPS receivers. Dual-frequency carrier-phase

Table 2.1: Real-time differential signals for GPS receivers

Increasing level of accuracy	Common differential correction sources
↑	A local base station
	Continuously operating reference station (CORS) network (e.g. GPSnet)
	Wide area (WADGPS) network (e.g. OmniStar VBS, XP, HP, Starfire SF1, SF2)
	Free-to-air coastal navigation beacon (Australian Marine Safety Authority)

receivers operating with a local base station correction are the most accurate DGPS available for civilian use (Table 2.1).

The expected accuracy from GPS receivers

Table 2.2 contains some general estimates that can be used as a guide to the average accuracy obtained with the different operational modes of receivers. The achieved accuracy of a specific receiver will depend on the contribution of the errors described above, at a given place and time.

Receiver accuracy measurement terms

Receivers usually have a stated accuracy level. In terms of stationary tests, the usual process is to record the calculated position from a receiver at a known point for a period of time and see how close the measurements are to the real location. Each recorded point will vary around the real location by some distance, and the way this variation is described causes some confusion in the agricultural community.

There is a range of standard reporting terms (Table 2.3). For example, when a receiver is reported to have a circular error probability (CEP) of ±2 m, then 50% of

Table 2.2: Average accuracy expected from different GPS operational modes

Operational mode		Average horizontal accuracy (RMS h)
Stand-alone	C/A code	6–10 m
	Carrier phase	3–6 m
Differential correction	C/A code	1–5 m
	Carrier-smoothed code (single carrier-phase)	0.1–1 m
	Dual carrier-phase (RTK)	1–5 cm
	Dual carrier-phase (static)	1–5 mm

Table 2.3: Accuracy reporting terms used by receiver manufacturers

Accuracy measures		Probability of a GPS position falling within this error range	Definition
CEP	Circular error probability	50%	The radius of circle centred at the true position, containing 50% of the position estimates in a scatter plot
R68	Horizontal 68	68%	The radius of circle centred at the true position, containing 68% of the position estimates in a scatter plot
R95	Horizontal 95	95%	The radius of circle centred at the true position, containing 95% of the position estimates in a scatter plot
RMS (horizontal)	Root mean square error	68%	The square root of the average of the squared horizontal position errors
2RMS (horizontal)	2 × root mean square error	95%	Twice the RMS of the horizontal position errors
1 sigma	1 standard deviation	68%	The standard error of the distribution of position estimates
2 sigma	2 standard deviations	95%	Two times the standard error of the distribution of position estimates
SEP	Spherical error probability	50%	The radius of sphere centred at the true position, containing the position estimate in 3D with probability of 50%
3D RMS	Three-dimensional root mean square error	68%	The square root of the average of the squared 3D position errors

the recorded data points can be expected to fall within 2 m of the real location while 50% are likely to be further away from the real location. As can be seen from the probabilities in Table 2.3, a number of commonly reported terms can be interchanged. Some very minor statistical licence has been taken, but for practical purposes the probabilities are applicable. A rule-of-thumb calculator for converting between the different accuracy terms is provided in Table 2.4. To convert from a quoted accuracy in terms listed in the rows, to accuracy in terms listed in the columns, read down the column to the appropriate row for the multiplication factor.

Agricultural uses for GPS

GNSS are becoming widespread in farm management and GPS receivers are by far the most common PA technology being adopted on-farm. This is mainly due to the tangible benefits associated with guidance, autosteer, control traffic farming (CTF)

Table 2.4: Conversion factors used to compare the accuracy in different terms. Multiply by the factor at the intersection of a column and a row to convert between accuracy terms

	CEP	R68	R95	RMS (h)	2RMS (h)	1 sigma	2 sigma	SEP	3D RMS
CEP	1.0	1.2	2.1	1.2	2.1	1.2	2.1	2.0	2.5
R68	0.83	1.0	2.0	1.0	2.0	1.0	2.0	1.7	2.1
R95	0.48	0.5	1.0	2.0	1.0	0.5	1.0	0.96	1.2
RMS (h)	0.83	1.0	0.5	1.0	2.0	1.0	2.0	1.7	2.1
2RMS (h)	0.48	0.5	1.0	0.5	1.0	0.5	1.0	0.96	1.2
1 sigma	0.83	1.0	2.0	1.0	2.0	1.0	2.0	1.7	2.1
2 sigma	0.48	0.5	1.0	0.5	1.0	0.5	1.0	0.96	1.2
SEP	0.5	0.59	1.04	0.59	1.04	0.59	1.04	1.0	1.27
3D RMS	0.4	0.48	0.83	0.48	0.83	0.48	0.83	0.79	1.0

and tramlining. Farmers can calculate and appreciate these benefits and are therefore quick to adopt them. The dollar benefits of other PA technologies are more site-specific or have a longer return on investment (see Chapter 8, 'Economics of PA'). Although guidance/autosteer is driving GPS adoption, receivers can be used for a variety of purposes (Table 2.5).

Different farm activities require different levels of positioning accuracy. While carrier-phase systems can be used for geo-referencing soil samples, cheaper and less accurate systems are just as useful as only metre-level accuracy is required. However, when centimetre accuracy is required, only carrier-phase RTK is applicable at present. RTK DGPS can be used for all PA operations currently performed in Australia.

Table 2.5: Possible applications for different GPS receiver operation modes in PA

	Stand-alone GPS	DGPS (C/A code)	DGPS (single carrier-phase)	Dual carrier-phase RTK (own base station or CORS network)
Soil and tissue tests	✓	✓	✓	✓
Crop scouting	✓	✓	✓	✓
Fertiliser strips	✓	✓	✓	✓
Strategic trials	✓	✓	✓	✓
Yield mapping	✓	✓	✓	✓
Variable-rate control	✓	✓	✓	✓
Vehicle guidance		✓	✓	✓
Elevation mapping			✓	✓
Land levelling and forming				✓
Vehicle autosteer				✓

Coordinate systems

A coordinate system provides a framework in which to identify and interpret locations with respect to an origin or reference point. For GNSS receivers there are two main types of coordinate systems – geographic and Cartesian – and locations can be converted between the two systems.

A geographic coordinate system identifies locations in latitude and longitude using degrees, minutes and seconds. The practical reference points for these systems are the equator (designated 0° latitude) and the Greenwich (or prime) meridian which is a line drawn between the two Poles and passing through Greenwich (designated 0° longitude).

The latitude of a specific location is the angle (measured from the centre of the Earth) that subtends an arc drawn from the specific location to the equator. The respective Poles are designated 90° North (positive latitude) and 90° South (negative latitude). The longitude of a location is the angle (measured from the centre of the Earth) that subtends an arc drawn at the equator from the prime meridian to where a meridian, passing from Pole to Pole and through the specific location, would intersect the equator. Longitude ranges from 0°–180° East (positive longitude) and 0°–180° West (negative longitude) from the prime meridian.

These geographic coordinate systems are fundamentally defined by geometrical models, known as spheroids, which are used to simplify the description of the Earth's surface. These descriptions are collectively known as datums. Each coordinate system employs a unique datum that affects the actual coordinates obtained for any particular location. It is important, when describing the geographic coordinates of a location, to state (or understand) which datum has been used.

The GPS uses a geographic coordinate system based on a datum known as World Geodetic System 1984 (WGS84). Australia uses a different datum, the Geocentric Datum of Australia 1994 (GDA94), to define the geographic coordinates of a location. However, for all practical purposes in PA, the two datums are the same and locations provided using the two datums are comparable.

Within each geographic coordinate system, each point on the planet has a unique latitude and longitude. However, an issue arises with geographic coordinates when the distance between locations needs to be calculated. While a single degree of latitude represents the same distance anywhere from the Poles to the equator, a single degree of longitude represents different distances, depending on the latitude. The extreme example is that at the equator, 360° of longitude is approximately 40 000 km, while at either Pole, 360° of longitude is a pirouette.

In this situation a Cartesian coordinate system, which has locations defined in metres from a defined origin, is needed. Converting from geographic to Cartesian coordinates involves a process of projecting the globe-like Earth onto a flat plane. The main Cartesian coordinate system in Australia is the Map Grid of Australia

(MGA94), which is based on the Universal Transverse Mercator (UTM) projection system which converts the angular measures of longitude and latitude into Eastings (E) and Northings (N).

To minimise distortion in the projection process, zones with a width equal to 6° longitude are projected from Pole to Pole, which divides the Earth into 60 zones. Each zone has its own origin from which the coordinates it encapsulates are calculated. The origin is the point at which the equator and the zone's central meridian cross. If the origin was denoted 0 E, 0 N then the southern half of each zone would be described by negative Northing coordinates and the western section of each zone would be described by negative Easting coordinates. To overcome this issue, a 'false origin' is created in each zone by designating the coordinates at the equator/central meridian intersection as 500 000.00 m E and 10 000 000.00 m N.

In practical terms, this means that Australia is covered by seven zones, numbered 50–56 from west to east. In each zone a location would be described by a six-figure Easting (approximately 250 000–750 000) and a seven-figure Northing (approximately 5 500 000–8 500 000).

In this system, coordinates repeat for each zone. So, while geographic coordinates are unique, any pair of MGA94/UTM coordinates by themselves can refer to six different locations throughout Australia. The zone number is therefore an important piece of information to be included with any Cartesian coordinate information.

In summary, it is always best to record and store original location data in geographic coordinates, and convert locations to Cartesian coordinates when an absolute spatial interpretation or analysis is required.

GNSS-based vehicle navigation systems

Advances in GNSS technology since the late 1990s have opened the door for the development of vehicle navigation guidance and autosteering systems for use on agricultural vehicles. These advanced navigation systems were pioneered in Australia and have been widely adopted for use in nearly all vehicle-based cropping operations.

Guidance systems

With this type of system, machinery control remains with the operator. It is the operator's responsibility to ensure that the machinery is steering in the intended direction. The guidance system uses a signalling device to prompt the driver to maintain a predetermined path. The operator can use personal judgement to override or correct any perceived errors associated with poor GPS positioning. Sub-metre accuracy DGPS is often used for this form of vehicle navigation assistance.

Autosteer systems

These remove the operator from the majority of steering operations. At present the majority of these systems require manual assistance at the end of each 'run' to direct the vehicle towards the beginning of the next 'run'. For safety, all autosteer systems have an automatic override system as soon as the operator takes control of the steering wheel. These systems also monitor the quality of the GPS signal and the autosteer will disengage (or not engage) if the GPS data is not of significant quality. Autosteer systems can be subdivided into two categories – steering assist and integrated autosteer.

Steering assist

This is an interim level between guidance and integrated autosteer systems. Assisted steering systems have an attachment to the steering wheel that allows the machine to be steered by manipulating the steering wheel. These systems are potentially less accurate than integrated autosteer systems given the nature of the steering control.

Integrated autosteer

The desired track information is passed directly to the vehicle's steering system through electronic control of in-line hydraulic valves. These systems require a steering kit to be fitted to each vehicle (unless it has already been factory-fitted) to allow communication/response between the autosteer computer and the hydraulic system. Using an RTK GPS with integrated autosteer is considered the most accurate option at present.

Benefit of vehicle navigation systems

The benefits of guidance/autosteer relate to the accuracy of the system being used and the management and quality of the driving that took place before installation. Listed from easiest to hardest to achieve, the main benefits are:

- reduced skip and overlap of inputs;
- improved timeliness;
- reduced driver fatigue;
- modifications to labour requirements;
- reduced compaction;
- improved soil water management;
- increased yield;
- precise seed-bed manipulation (e.g. raised beds);
- inter-row cultivation and spraying;
- inter-row planting.

Integrated autosteer with positioning provided by RTK GPS offers the best chance of maximising the degree of benefit gained and is certainly required to achieve inter-row applications.

Key points

- Global navigation satellite system (GNSS) is the industry standard generic term for satellite-based systems that provide autonomous geo-spatial positioning with a global coverage. See http://www.ignss.org/.
- The Navstar global positioning system (GPS) is one type of GNSS. It is controlled by the US Department of Defence. Russia, in partnership with India, is restoring the Soviet-built Global'naya Navigatsionnaya Sputnikovaya Sistema (GLONASS) to operational capacity. A European consortium will have GALILEO operational by 2013. These are or will be free signals.
- There are three main segments to a GNSS:
  - the space segment (a constellation of orbiting satellites);
  - the user segment (users equipped with receiver units);
  - the control segment (master control station and satellite-monitoring ground stations).
- GNSS receivers need line-of-sight to at least four different satellites to be able to triangulate their position using the GNSS signals.
- There are three different types of receiver operation with different operational accuracy levels. Accuracy comes at a price:
  - stand-alone receivers: approximately 10 m accuracy, A$100–500;
  - subscription-based differential receivers: 0.1–1 m accuracy, up to A$10 000;
  - carrier-phase real-time kinematic (RTK) differential receivers: 2–10 cm accuracy, A$10 000–40 000.
- The accuracy of the different receivers is dependent on how accurately the distance from each satellite to the receiver can be measured.
- The main causes of positioning failure are:
  - power failure;
  - blocked line-of-sight;
  - electrical or magnetic interference.
- A differential signal can be used to help correct the errors in the basic GNSS measurement.
- A differential signal is not provided by the GNSS. It can be provided by private companies, government agencies or private base stations.
- Receivers often take time to warm up. This is known as the acquisition time and its length depends on where and when the receiver last operated.
- Vehicle navigation assistance using GPS requires at least subscription-based differential receivers.
- Highly accurate vehicle autosteer requires carrier-phase RTK receivers.
- All GNSS use latitude and longitude (measured in degrees, minutes and seconds) to describe a receiver's location. Latitude and longitude can be converted into Eastings and Northings (measured in metres) to make distance calculations easier.

3

Hardware for Precision Agriculture

The major agricultural machinery brands actively support PA and often include PA hardware systems as standard features on new equipment. Hardware components such as variable-rate controllers and yield monitors are also available from third-party suppliers. Other crop and soil sensing technologies can be easily accessed through consultants or service providers.

Yield monitoring systems

Crop yield can be automatically measured and recorded on a harvester during the harvest process. The grains industry pioneered the use of these systems, due largely to the extensive use of modern machinery and the suitability of grain harvesters for adaptation. A grain yield monitoring system requires a number of sensors (Figure 3.1) and some computer equipment to be installed on the harvester.

For yield monitoring and mapping, sensors are needed to measure:

- clean-grain flow;
- grain moisture content;
- grain (elevator) speed;
- cutting comb height;
- harvester ground speed;
- harvester location.

The information gathered by these sensors is transmitted to a cabin-mounted monitor that combines a basic computer with recording and display software and a memory port for data storage and transfer to an office or laptop computer.

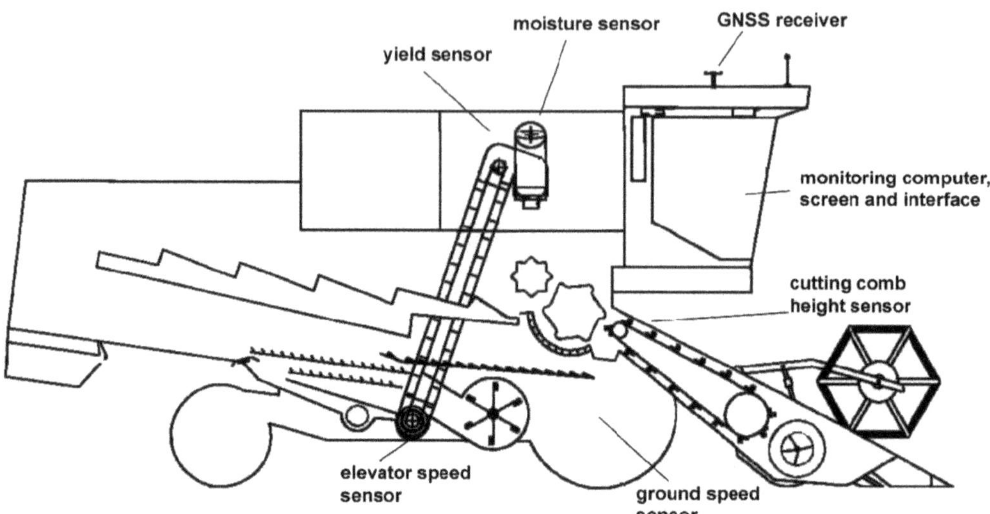

Figure 3.1: General locations of yield monitoring components on a harvester.

Components of a yield monitoring system

Grain flow sensors

The two most common types of system used in Australia involve different ways of measuring grain flow inside the harvester.

The impact-plate type calculates a mass flow directly from measurements of the force of grain hitting a plate connected to a strain gauge or potentiometer. This type of sensor is usually located at the top of the clean-grain elevator to intercept the grain before it reaches the grain-bin auger. Calculating the mass this way requires information on the speed at which the clean-grain elevator is running.

The volumetric-type sensor measures the volume of grain on individual elevator paddles as they rise past a light-beam that is directed across the interior of the elevator. The measured grain volume is combined with a grain-specific weight to provide an estimate of mass. Determining the mass this way requires measurement of any significant deviation of the harvester from the level.

These two types of grain yield sensor cannot differentiate between grain seed and foreign matter.

Grain moisture sensors

Determining moisture content of the grain is important for various reasons, including when to harvest, and for storage, handling and marketing decisions. Several types of grain moisture sensors are commercially available. The most common rely on the correlation between the moisture content of grain and its electrical properties. These systems measure the capacitance of the grain at a position in the clean-grain flow, close to the grain flow sensor. All the major brands of yield monitor use this system, in slightly different physical configurations and locations in the grain flow. These types of moisture sensors can be susceptible to surface moisture on the grain, so fluctuations in recorded grain moisture can be exaggerated in the early morning and evening under certain conditions.

Grain moisture can also be monitored using near-infrared (NIR) technologies. These sensors rely on the relationship between grain moisture content and its effect on the transmittance or reflection of near-infrared wavelengths. NIR measurement has been used in grain receival silos for many years. On-harvester application of NIR for moisture measurement is now being combined with grain quality sensors. The use of the NIR relationship is much less susceptible to moisture on the grain surface.

Elevator speed sensors

A speed sensor is necessary to establish the speed of the clean-grain elevator. For the impact-plate system, which uses direct mass flow measurement, the speed at which the grain leaves the paddle at the top of the elevator is used to calculate its acceleration towards the impact plate. The impact plate registers a force which, when divided by the acceleration of the grain, provides a figure for the mass of grain. In volumetric systems, the speed is used to provide a flow rate to combine with the mass estimate produced by the volume multiplied by specific mass calculations.

Comb height sensors

The signal to the yield monitoring system that the harvest process has started or stopped is controlled by a sensor that registers the height of the cutting comb. When the comb is lowered within normal operating range, a signal from the sensor initiates recording of data. When the comb is raised above the normal cutting height range, an 'up' signal stops data recording. The type of sensor varies with manufacturer.

Ground speed sensors

Ground speed can be measured by tapping into existing axle/transmission rotation sensors, using radar sensors or, with increasing accuracy, global navigation satellite system (GNSS) data. Different brands and models of yield monitors can use one or all of these approaches. The ground speed (metres per second) is used to calculate the distance the harvester has travelled (metres) over the desired measurement period (seconds).

GNSS

Connecting these systems to a GNSS enables the monitoring system to become a mapping system. At each desired time interval, the data collected from the other sensors is matched with a location retrieved from data streaming from the GNSS receiver. All GNSS receivers on the Australian market can provide position updates at 1 second intervals; many of the more accurate models can supply information up to 10 times per second (10 Hz).

Yield monitor console

Mounted in the cabin of the harvester, yield monitor consoles are now essentially tablet computers comprising a central processing unit, screen, and data input and storage features. These units use proprietary operating systems. They collect, link by time, and store the data from the individual sensors that is necessary to calculate the crop yield per hectare. Operational aspects of each brand of yield monitoring systems differ but all allow the input of field information, calibration functions and visual display of the calculated yield and moisture at the sampling interval requested. Most modern systems also provide additional information on the display such as grain throughput (tonnes per hour).

Additional sensors for vehicle orientation

The volumetric-type flow sensors rely on calibrating grain volume to the period of time a light beam shone across the grain elevator is broken as a grain paddle passes. The way the pile of grain sits on each paddle is generally constant if the harvester is always level. If the harvester is subject to significant slope forward or backward (pitch) or sideways (roll), the grain on the paddles may sit in a manner that makes the sensor over- or underestimate the amount of grain. Pitch issues can also affect impact-plate systems, as operating on significant up- or down-slopes can bias the recorded force of the grain flow.

A system of electronic gyroscopes is usually used to gather information on the orientation of the harvester, to compensate in the estimation of grain flow. These sensors should always be deployed with volumetric-flow sensors, but are considered a necessary addition to impact-type sensors only if the harvester is to be operated on significant slopes. In new harvesters, the orientation sensors are often fitted as components of autosteer instrumentation and so are becoming increasingly utilised.

Information recorded by yield monitoring systems

Each yield monitoring system records sensor data in a proprietary, binary form. The data is saved to file on a recordable data card/media, either directly during harvest operations or when a the data is downloaded from the harvester. The data files must be opened into the software provided by the equipment manufacturer, third-party mapping software or an external data viewer that accepts the specific format. The data recorded in these files will have information similar to the spreadsheet file displayed in Table 3.1.

All yield monitors record enough information to calculate the instantaneous yield at each measurement location as well as the grain moisture content, GNSS elevation and information about the GNSS signal, crop type and field.

Table 3.1: Yield monitor information showing the type of data recorded at 1 second (1 Hz) intervals during the harvest operation

Longitude	Latitude	Flow (kg/s)	GPS time	Measurement cycle (s)	Distance travelled (mm)	Swath width (mm)	Moisture (%)	Satellite status	Pass number	Harvester serial number	Field ID	Load ID	Crop type	GPS status	Elevation (m)
150.003681	-29.829868	6.35	973562564	1	2440	8990	10.8	33	1	981673	"F7"	"L2"	"Wheat"	7	310.1
150.003706	-29.829871	6.21	973562565	1	2440	8990	10.7	33	1	981673	"F7"	"L2"	"Wheat"	7	310.3
150.00373	-29.829878	6.06	973562566	1	2460	8990	10.9	33	1	981673	"F7"	"L2"	"Wheat"	7	310.3
150.003755	-29.829883	6.63	973562567	1	2510	8990	10.9	33	1	981673	"F7"	"L2"	"Wheat"	7	310.4
150.00378	-29.829886	6.49	973562568	1	2510	8990	10.8	33	1	981673	"F7"	"L2"	"Wheat"	7	310.3
150.003805	-29.829893	6.54	973562569	1	2510	8990	10.9	33	1	981673	"F7"	"L2"	"Wheat"	7	310.3
150.003823	-29.829898	6.54	973562570	1	2510	8990	10.9	33	1	981673	"F7"	"L2"	"Wheat"	7	312.1
150.00385	-29.829905	5.86	973562571	1	2490	8990	10.7	33	1	981673	"F7"	"L2"	"Wheat"	7	311.4
150.003875	-29.82991	6.96	973562572	1	2460	8990	10.7	33	1	981673	"F7"	"L2"	"Wheat"	7	311.4
150.003898	-29.829916	5.72	973562573	1	2490	8990	10.9	33	1	981673	"F7"	"L2"	"Wheat"	7	311.6
150.003923	-29.82992	6.48	973562574	1	2460	8990	10.8	33	1	981673	"F7"	"L2"	"Wheat"	7	311.6
150.003948	-29.829925	6.3	973562575	1	2540	8990	10.9	33	1	981673	"F7"	"L2"	"Wheat"	7	311.7
150.003973	-29.829928	6.39	973562576	1	2510	8990	11.1	33	1	981673	"F7"	"L2"	"Wheat"	7	311.7
150.003998	-29.829935	6.89	973562577	1	2540	8990	11.1	33	1	981673	"F7"	"L2"	"Wheat"	7	311.7

Calculating crop yield

There are a number of different approaches to physically monitoring the yield on a harvester. They all provide a final output of crop yield in tonnes per hectare (t/ha) at a user-set time interval, usually 1 second (1 Hz), when the harvester is operating within a field. To provide this information, it is necessary to gather data on the mass of the crop being harvested and the area covered during the time period.

The mass of the crop is determined using a grain flow sensor in combination with data on the speed of clean-grain flow within the harvester. The area covered is determined by the width of the harvester comb and the distance the harvester travelled in the time period. The yield can be corrected to a constant moisture content if the actual grain moisture content is recorded at the same time.

$$Y_{t/ha} = \left(\left(\frac{f_{kg/s} \times i_s}{1000_{kg/s}} \right) \times \left(\frac{10\,000_{m^2/ha}}{d_m \times w_m} \right) \right) \times \left(\frac{100 - m_{\%}}{100 - sm_{\%}} \right) \qquad \text{(Eq. 3.1)}$$

where:

$Y_{t/ha}$ = crop yield in t/ha

$f_{kg/s}$ = mass of grain flow in kg/s

i_s = sample interval (s)

d_m = distance travelled (m)

w_m = width of the harvester comb (m)

$m_{\%}$ = moisture content of harvested grain (%)

$sm_{\%}$ = set moisture content for market reference (%).

Yield calculated at every measurement location produces data such as shown in Figure 3.2. Extensive within-field variation is evident, with yield ranging from 1.2 t/ha to 5.5 t/ha. The boxes a, b and c enclose 1 ha areas which are uniform/low, uniform/high and variable/medium respectively. Given that input management was uniform across the field, this data highlights the significant differences in production and expected profit. Such data is driving the exploration and development of SSCM.

Yield monitor availability and calibration

All new commercial grain harvesters can be purchased with factory-installed yield monitors. After-market yield monitors are available for installation on many late-model harvesters.

The yield monitor comes with factory calibrations, but to gather accurate yield data it is necessary to perform a local calibration for each type of grain harvested. Calibration involves harvesting a number of loads of grain and being able to

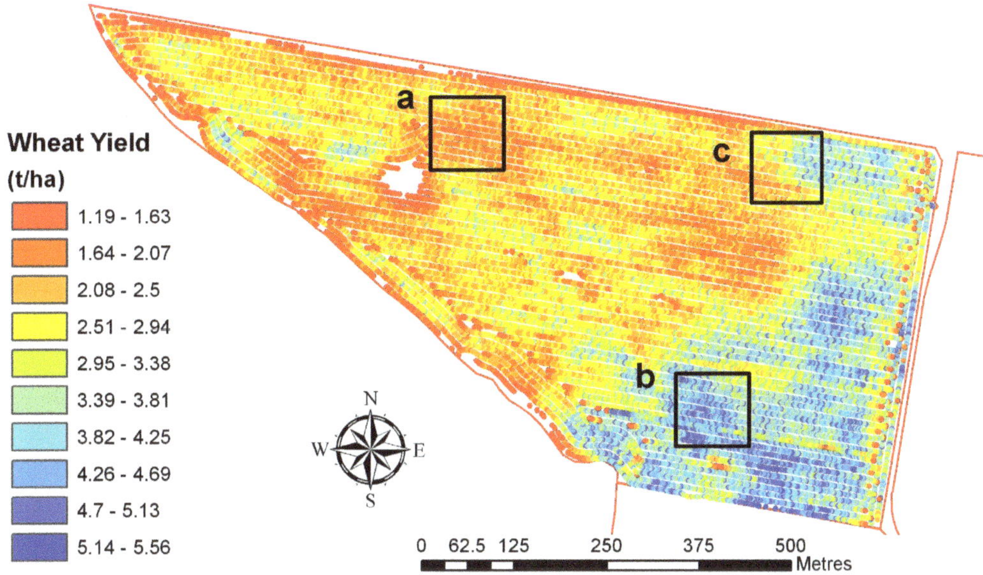

Figure 3.2: Yield data recorded at 1 second (1 Hz) intervals in a 40 ha field.

record the actual weight and moisture content. This information is fed into the yield monitor console and is used to automatically adjust the on-board yield calculation formula.

The manufacturer's handbook should be consulted for brand-specific calibration instructions.

Operational issues

The major operational issue with these monitors is matching location information with the other sensor data. At any point when the grain flow is recorded at the sensor, the grain that is being measured has been cut from the crop, passed through the threshing, separating and cross-augering process to be presented to the flow sensor.

Clearly, the grain was harvested from a location that the harvester had traversed earlier. This 'transport delay' in presenting harvested grain to the flow sensor is compensated for by matching the sensor measurements to location information collected at an earlier point in time. Most brands of yield monitor allow this interval to be adjusted (often between 5 and 15 seconds). It is possible to determine the size of the transport delay for a particular harvester operating at a standard speed and harvest capacity.

It is also important to understand that the quality of the yield data recorded by the system depends on the calibration and the actual operation of the harvester. Data quality is improved when:

- harvesting is performed in straight lines;
- navigation aids such as guidance or autosteer are used;
- an even flow of material into the harvester is maintained.

Data quality may decline when:

- frequent stopping or erratic changes in speed occur;
- harvesting with a partially full comb;
- travelling with the comb down while not harvesting.

Grain protein and oil content monitors

Profits from grain crops are not derived only from yield; the quality of the grain delivered is critical to ensuring maximum profits for some grains. For cereals, the quality is largely determined by the grain protein content. For canola, corn and soybeans the oil content is important.

Protein and oil content is determined by:

- grain crop type;
- crop variety;
- nitrogen in the soil and applied as fertiliser;
- moisture availability during the growing season.

Accurately measuring grain protein or oil content in conjunction with yield is useful in assessing how well crop nitrogen nutrition was managed in the harvested season and for locating areas where management may be altered in the future.

Grain quality monitors can be mounted on a harvester to measure protein and oil content during harvest operations. These sensors rely on the use of near-infrared spectroscopy (NIRS), the same technology employed in grain receival depots. Grain samples are presented to a light source; the light interacts with specific chemical bonds within the grain and is slightly modified because the bonds absorb some of the light energy. The reflected or transmitted light (depending on sensor brand) is measured and this provides a surrogate measure of grain protein, moisture and oil content.

Instruments that use near-infrared transmittance (NIT) need to capture a stationary sample, so the measurement readings are obtained every 10–15 seconds (a measurement cycle of 0.1–0.07 Hz). Using near-infrared reflectance (NIR), a continuous grain flow could be used and the measurement taken more frequently.

Within-field variation in grain protein content, like grain yield, appears to be widespread (Figure 3.3). This data was obtained during the same harvesting operation as the yield data in Figure 3.2. The difference in sampling intensity between the yield (measured at 1 Hz) and the protein (measured at 0.08 Hz) produces a difference in observation density (yield = 725 obs/ha and protein = 65 obs/ha).

Using grain protein content in PA

While yield maps document variation in the mass of output, accurately measuring variability in grain protein content across a field may provide a range of other benefits.

Nitrogen agronomy

Conditions that might increase the range of grain protein content found within a field include:

- variation in nitrogen availability within-field. Where different soil type or soil texture occurs, there may be variation in nitrogen supplies to the crop;
- variation in moisture availability within-field. The interaction between topography, soil type and seasonal climatic conditions will result in spatial variation in the amount of water available to a crop across a field. This impacts on nutrient uptake, grain filling and yield, which all control final grain protein content.

Locations in a field where grain protein is identified as lower provide an opportunity to direct investigations into nitrogen availability and inform any options for management intervention. If investigations suggest that a field has received adequate nitrogen for the growing season then, in the simplest situation, a

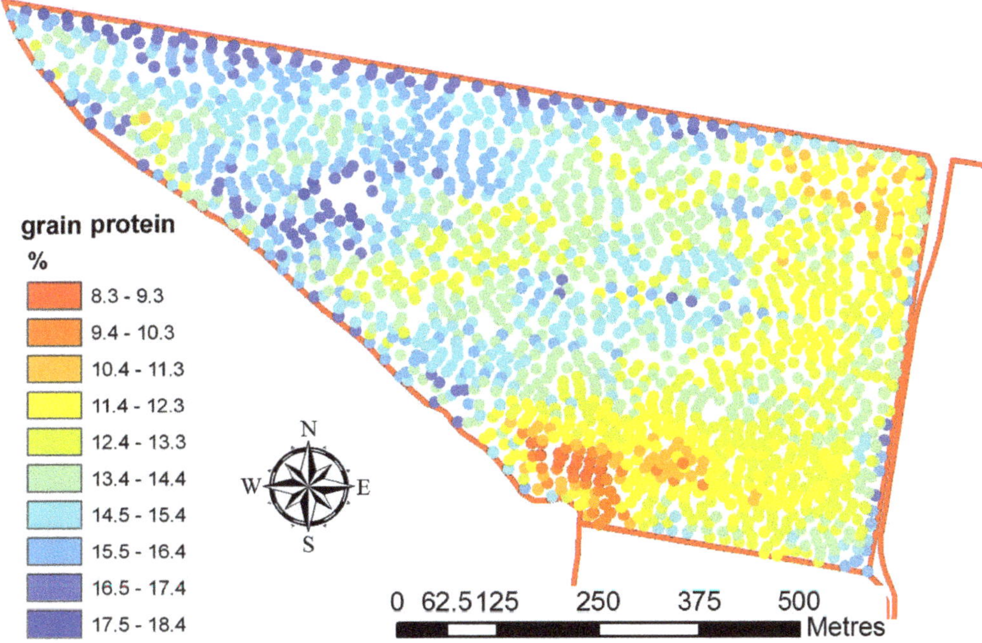

Figure 3.3: Protein data recorded at 12 second (0.08 Hz) intervals in a 40 ha field.

nitrogen-removed map may become an option to guide variable-rate nitrogen replacement for the following season. This is because grain protein is the main mechanism for the movement of nitrogen off-farm.

The amount of nitrogen removed from the production system through harvested grain can be calculated by multiplying the mass of grain yield by the percentage of protein in the grain.

Premium quality marketing and gross margin maps

Identifying areas within a field or farm where grain with specific protein levels can be harvested may provide premium marketing and increased profit opportunities.

Growers who are able to provide grain at a consistent quality may be in a better position to obtain premium contracts. In highly variable years, the ability to segregate part of the crop based on protein percentage may allow growers to meet contracts when the average quality of the farm output fails to meet the contract specifications (provided that the entire crop is not under contract). In these scenarios, a protein sensor opens the possibility of differential harvesting to target grain quality.

It is important to remember that quality monitoring can be done at any stage of the supply chain and should be performed regularly to assure quality levels and to maximise market segregation opportunities. NIRS quality monitors have been used successfully off-harvester to segregate or manage grain for delivery, based on protein content. This does not provide spatial information on protein variability within fields but it can be very effective as a differential marketing technique. The malting barley market is one example in Australia where off-harvester segregation is being used to maintain delivered product within a contracted quality window. To achieve this, the protein sensor is mounted on a silo auger and the output controls a mechanism that diverts the grain into different silos depending on protein content.

In the Australian wheat and malting barley markets, premiums and discounts are applied to the sale price based on the quality and moisture content of delivered grain. With yield, moisture and protein maps available, it is therefore possible to calculate more accurate revenue figures and produce maps of how the gross margin of production varies within a field and across a farm. These maps would be very useful to a farm manager, in particular for planning profitable crop rotations or identifying repeatedly unprofitable areas for alternate uses.

Calibration and maintenance of quality sensors

Factory calibrations are supplied with quality sensors, however, a local calibration check is advisable. A number of grain samples can be analysed at the local receival silo then passed through the harvester sensing system. It is wise to check calibrations for all crop types and varieties to be harvested. Dust and material other than grain will affect the operation of quality sensors, and efforts should be made to keep the grain sample clean.

Soil sensing systems

A yield or protein map can show the variability of crop performance within a field. However, these maps give little indication about the reasons for differences in crop performance. In some cases, it is related to management practices or past history. However, areas of lower performance are frequently related to factors associated with non-optimal soil conditions. A better understanding of the spatial patterns in soil properties that cause yield/quality variation can be achieved with maps of soil variability.

Traditionally, geo-referenced manual soil sampling and laboratory analysis has been available through commercial vendors. This normally involves sampling at only a small number of sites in the field, due to the labour intensiveness and financial cost of the procedure. Small sample numbers make it difficult to gain a true picture of variability in soil properties across a field. This in turn makes it hard to quantify any relationship between the soil properties and crop yield and quality variation, as seen in Figures 3.2 and 3.3.

One method of overcoming this problem is 'grid sampling', whereby the soil is sampled in a grid pattern within the field. A grid size of 1 ha has been used by some PA practitioners, particularly in the US. However, soil sampling for a comprehensive analysis at this scale would be cost-prohibitive in Australia at present. And even if expense wasn't an issue, the size of the grid would still be too coarse to create reasonably accurate maps of the variability in all the soil properties of agronomic interest using a single sample set. The sampling grid size would need to be closer to 0.5 ha to achieve the desired result, and this would certainly be uneconomical using traditional sampling and analytical techniques.

Being able to measure soil properties at many locations in a field using sensors mounted on a vehicle would increase the sampling rate and enable accurate maps to be constructed. Several research groups, as well as commercial institutions, have been investigating various proximal and invasive measurement techniques that could be suited to automated soil mapping (similar to yield monitoring). Instruments using the following measurement techniques are commercially available for on-the-go soil measurement.

Apparent soil electrical conductivity (EC_a) sensors

EC_a is a measure of the ability of the soil to conduct electricity. An electrical current may be conducted through the soil via three pathways:

- the pore-connected soil solution of water and ions;
- the cations that are bound to the surfaces of clay particles;
- connected solid soil particles.

Because of these three pathways, there are a number of soil attributes that will affect the soil's ability to conduct electricity. These are:

- soil texture (in particular, amount of clay);
- soil moisture;
- soil cation exchange capacity (CEC), driven by the type of clay minerals and soil organic matter;
- ions in the soil solution (dissolved salts which include fertilisers);
- soil temperature.

The higher the level of each attribute, the greater the electrical conductivity of the soil. Higher electrical conductivity generally means better fertility and yield potential. However, excessively high readings tend to indicate salinity. EC_a can be measured using two different techniques: electromagnetic induction (EMI) where the instrument can operate without touching the soil surface (a proximal sensor), or electrical resistivity (ER) instruments that need to penetrate the soil to work (an invasive sensor).

Electromagnetic induction

EMI instruments use paired transmitting and receiving induction coils at opposite ends of the instrument. The transmitting coil generates a primary magnetic field that induces eddy currents (swirling electrical currents) in the soil. These eddy currents induce a secondary magnetic field that is detected by the receiving coil. The strength of the secondary magnetic field is proportional to the amount of current flowing in the soil, from which the EC_a can be calculated.

The theoretical depth to which the current penetrates the soil is determined by the distance between the transmitting and receiving coils, the orientation of the coil dipoles, the frequency of the current and the conductivity of the material. Greater separation between the coils increases the depth of penetration. When the coil dipoles are horizontal, the depth of penetration is less than in the vertical mode. Penetration decreases as current frequency increases.

These functionalities have been combined in devices with different coil-pair arrays and combinations to provide a range of instruments with different nominal depths of exploration (DOE). The DOE is a practical estimate of the depth to which the EC_a is being measured by an array.

Table 3.2 lists a range of commercially available EMI instruments that are suitable for vehicle-mounted or towed surveys. Two of the most commonly used in PA are the Geonics 38DD and the DUALEM-21 (Figure 3.4). The output of all the listed instruments is an average of the EC_a from the soil surface to the stated DOE of an array. Instruments that list more than one DOE can provide EC_a measurements to the different depths simultaneously.

Horizontal and vertical dipoles

The Geonics EM38-DD configuration (Figure 3.4a) is used here as an example to explain the differences between the vertical and horizontal modes of operation. All

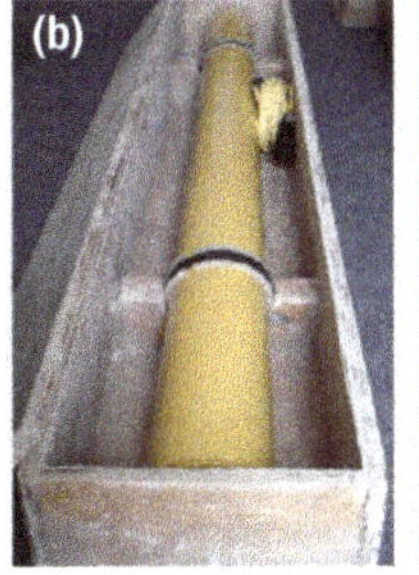

Figure 3.4: Two of the most common EMI instruments used in PA. (a) A Geonics 38DD (image courtesy of Geonics Ltd). (b) A DUALEM-21 mounted in sled. (c) A DUALEM-21 operating during an in-field survey (images b and c courtesy of Precision Agronomics Australia).

the instruments listed in Table 3.2 operate on the same general principle. The relative response curves for each mode (Figure 3.5) show that in the vertical dipole mode, more of the response is obtained from deeper in the soil profile, while in the horizontal mode, more response is obtained from higher in the soil profile. For this example, the vertical DOE = 1.5 m and horizontal DOE = 0.75 m.

Electrical resistivity instruments

An electrical resistivity (ER) instrument is an invasive soil sensor which operates by passing a current into the soil through an active metal electrode that must be in physical contact with the soil. The current is at a known voltage. A non-active

Table 3.2: EMI instruments available in Australia have coils at various spacings, frequencies and dipole orientations to give a broad range of exploration depths

Brand	Instrument	Depth of exploration (m)
Geonics Ltd	EM38 horizontal	0.75
	EM38 vertical	1.5
	EM38-DD	0.75 & 1.5
	EM38-MK2 horizontal	0.375 & 0.75
	EM38-MK2 vertical	0. 75 & 1.5
	EM31-MK2	6.0
	EM31-SH	4.0
DUALEM	DUALEM-1	0.5 & 1.6
	DUALEM-2	1.0 & 3.2
	DUALEM-4	2.0 & 6.4
	DUALEM-6	3.0 & 9.5
	DUALEM-21	0.5 & 1.0 & 1.6 & 3.2
	DUALEM-42	1.0 & 2.0 & 3.2 & 6.4
	DUALEM-421	0.5 & 1.0 & 1.6 & 2.0 & 3.2 & 6.4
	DUALEM-642	1.0 & 2.0 & 3.0 & 3.2 & 6.4 & 6.5

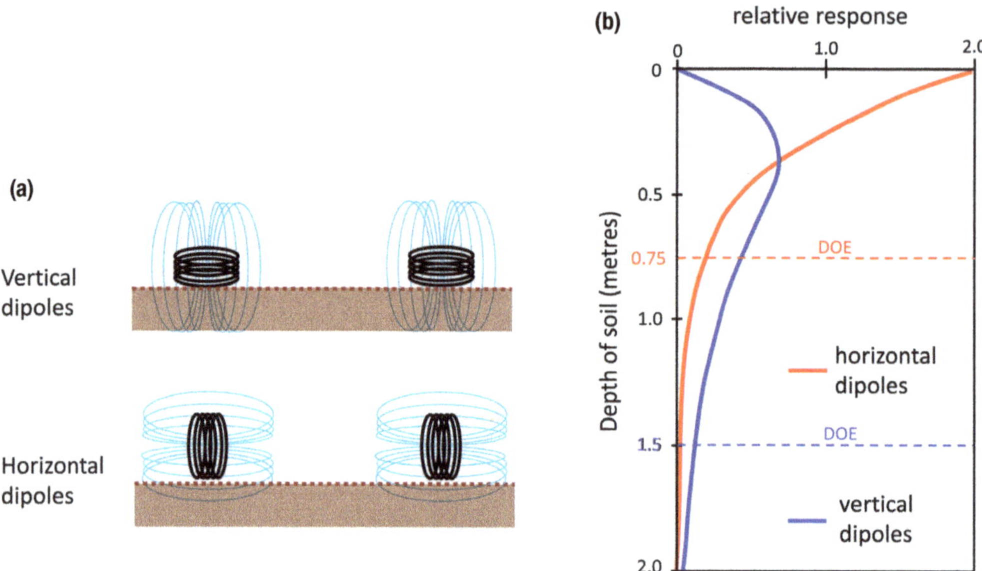

Figure 3.5: (a) A depiction of vertical and horizontal dipole orientations. (b) The effect on DOE for the EM38-DD EMI instrument.

electrode that is also in contact with the soil then measures the drop in voltage as the current travels through the soil. The drop in voltage can be related to the resistivity of the soil. Resistivity is inversely related to conductivity and from this the EC_a can be calculated. The distance between the two electrodes determines the DOE.

ER has been employed in a number of mobile instruments that use cultivation discs as the metal electrodes (Figure 3.6). The only commercial option in Australia is the Veris 3100. It uses two central discs as the active electrodes and two pairs of non-active disc electrodes set at two different distances from the active discs. The disc spacing can be altered, but the standard configuration provides two simultaneous DOEs of 0.3 m and 0.9 m. Using the knowledge of the instrument's depth response for the two DOEs, it is possible to calculate the contribution from the 0.3 m to 0.9 m soil layer and provide estimates of EC_a in three regions of a soil profile.

Output from the instruments

Figure 3.7a shows the raw EC_a measurements from an EM38h survey on a 27 m swath. Figure 3.7b and c show EC_a maps made from the EM38h (0–75 cm) survey data and that obtained from the same field using the Veris 3100 (0–90 cm) respectively. There is an obviously strong correlation between the patterns of variability, with the slightly greater DOE of the Veris 3100 pointing towards increasing EC_a with depth across much of this field. The crop yield map (Figure 3.7d) shows that the relationships between EC_a and crop yield can also be quite strong.

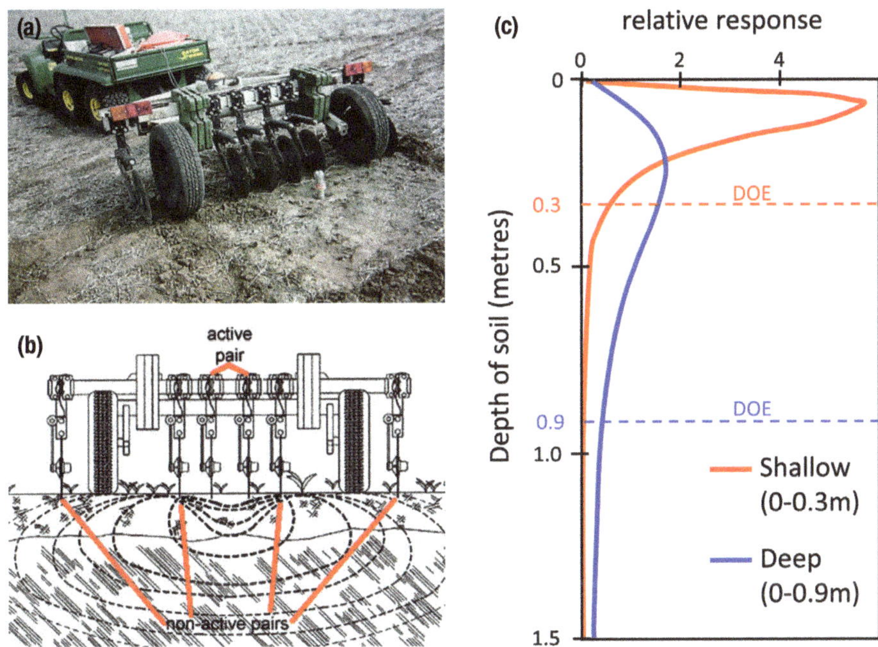

Figure 3.6: (a) The Veris 3100 ER instrument. (b) A schematic of its operation. (c) The DOE for the two non-active electrode pair spacings.

The investigative ability provided by these tools is further highlighted by a comparison of the maps obtained from different configurations of the EMI instruments. In Figure 3.8 the maps use a single legend scale, making it easy to see that the EC_a is increasing and becoming more variable with depth. The pattern builds consistently as the exploration goes deeper, but the shallow exploration shows that the topsoil is generally contributing less to the spatial pattern of variability in EC_a. Being able to 'see' into the soil at different depths helps to direct the location of soil sampling points to answer specific questions about what is driving the variability and the potential impact on crop production.

Using EC_a in PA

Soil properties that influence conductivity can be correlated to EC_a measurements by direct sampling and comparison. For agricultural applications, the most common primary soil variables that correlate well with EC_a are soil clay content, soil water content and soil salt content. However, the fact that these soil properties have an interchangeable effect on the EC_a makes a universal calibration for any one of the properties very difficult. In addition, temperature (air and soil), the mineral content and the clay type will affect any calibration. That said, with some effort it is possible to build calibrations for these properties within a field or even a farm.

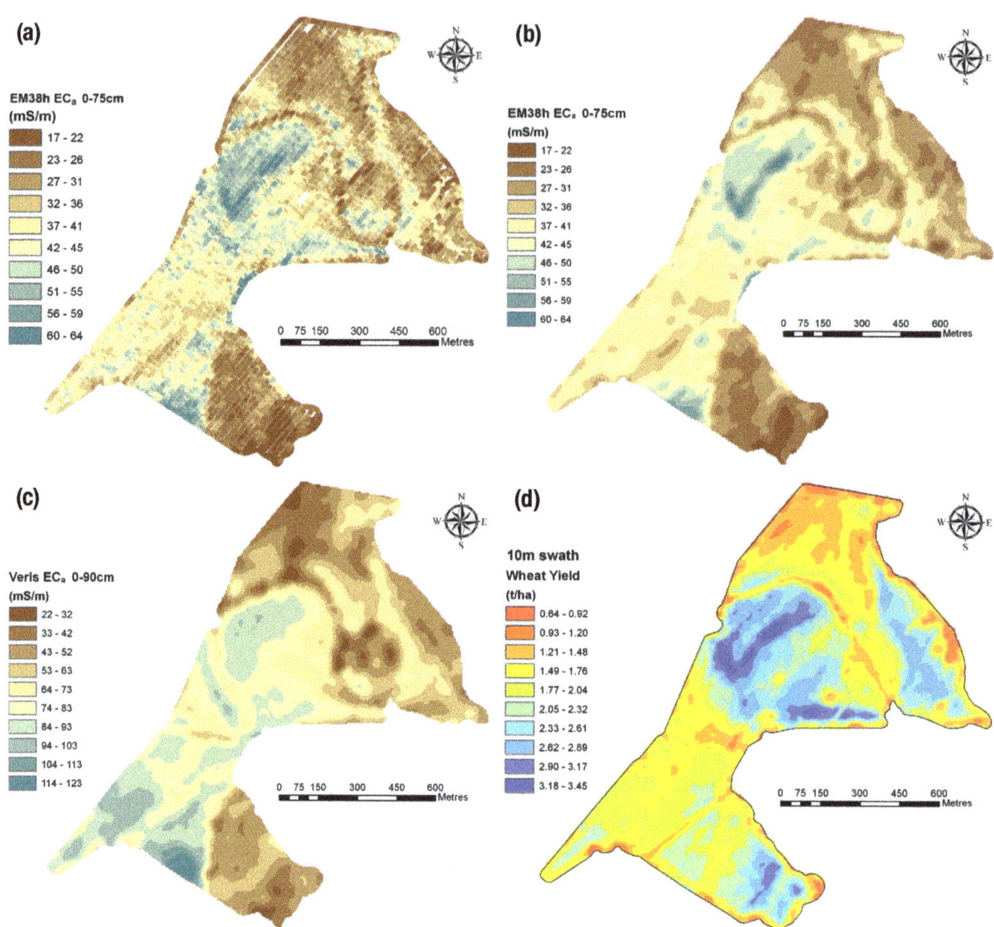

Figure 3.7: (a) Raw EC_a data gathered using an EM38h. (b) An EC_a map made from the EM38h (0–75 cm) data. (c) An EC_a map made from Veris 3100 (0–90 cm) data. (d) The subsequent wheat yield map.

Other indirect relationships can be established with EC_a at the field or possibly farm level. Depth of soil, cation exchange capacity (CEC), rooting depth, available water-holding capacity and physical and chemical constraints will all require customised calibrations for a site. The success of these calibrations is often determined by the type of soil at the location.

What is known is that the pattern of EC_a variation in a field usually reflects changes in the yield potential. This is best seen if the measurements have been taken when the soil moisture profile is reasonably full. This is because as the clay content increases, the soil's ability to store moisture also increases, as does the CEC (which increases the ability to store nutrients). Consequently, areas of higher EC_a readings within a field should have higher yield potential. If the EC_a measurements are taken after harvest, when the soil profile should be reasonably dry, it is possible

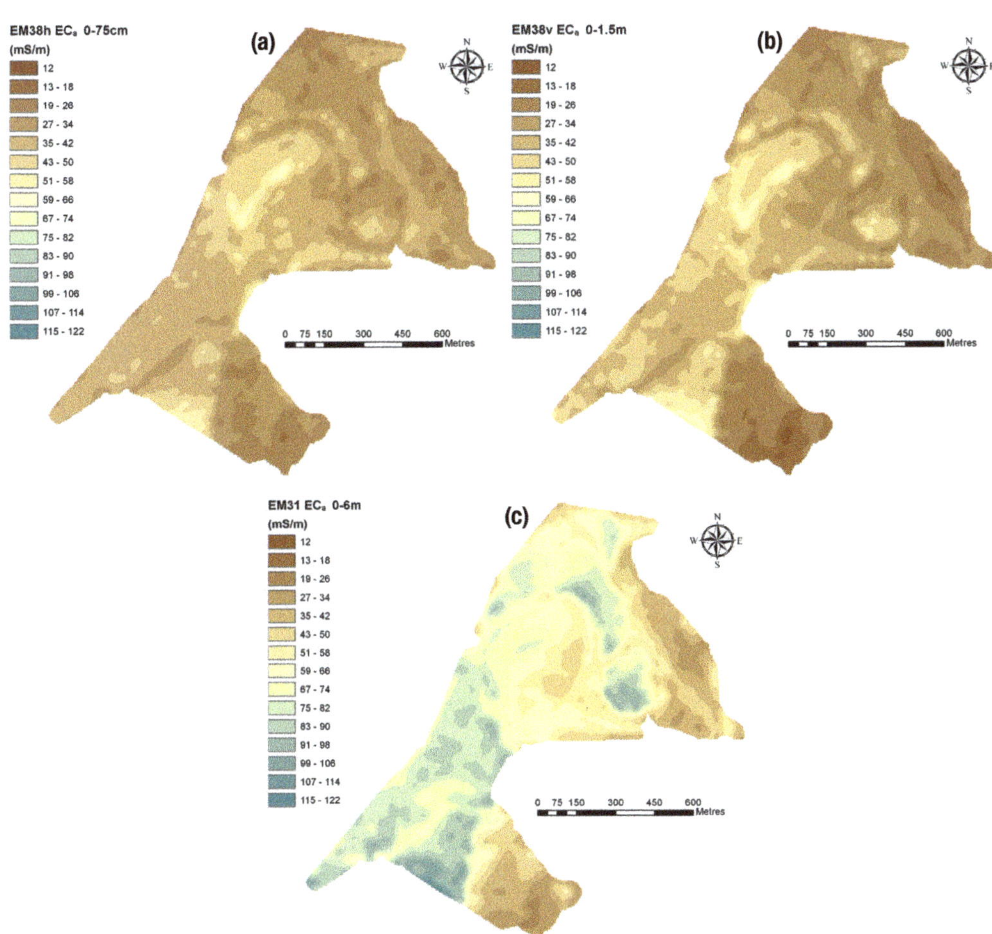

Figure 3.8: Comparison of soil EC_a measured by three different configurations of EMI instruments resulting in different DOEs. (a) EM38h (0–75 cm). (b) EM38v (0–1.5 m). (c) EM31-MK2 (0–6 m).

to calibrate the instrument to identify areas where soil moisture remains unused in the soil. These may be indicative of subsoil constraints.

However, at the moment, the most common use of soil EC_a maps remains the identification of soil sampling sites to explore the reasons for variability in crop production. The EC_a maps are used either by themselves or in conjunction with yield, elevation and remotely sensed images to pinpoint areas of production difference.

Gamma radiometer

Gamma radiometrics is the measurement of natural gamma ray emissions primarily from the top 40 cm of soil or rock. This can often provide information about the parent material of the soil that can be related to soil types across the region or field.

Gamma rays are emitted as high-energy short wavelength electromagnetic radiation and are part of the natural radioactive decay process in which alpha and beta particles and gamma rays are emitted. Unlike alpha and beta particles, gamma rays are detectable by remote sensors because they can travel a reasonable distance in air.

Measuring gamma radiation

Potassium (K), uranium (U) and thorium (Th) are the three major elements in soil that have naturally occurring isotopes which emit gamma rays as they decay. These can be measured in different parts of the energy spectrum. It is also usual to measure the total count (TC) across a specified energy spectrum range.

These measurements are performed using a scintillator, which is made from sodium or caesium iodide crystals containing a small amount of thallium. A photomultiplier is associated with the scintillator, so that the energy absorbed from any gamma rays can be registered and recorded as a voltage.

Attenuation

Fifty percent of the gamma rays that can be detected above the soil surface have been emitted from the top 10 cm of soil; 90% are from the top 40 cm of soil. Gamma rays from soil below 50 cm depth are blocked by the soil above. Gamma rays are also blocked by increasing soil moisture in the topsoil. A 1% increase in soil moisture content equates to a 1% increase in attenuation.

Standing water on the soil surface can block gamma ray detection, but vegetation cover has little impact unless measurements are being made over dense wet forests.

Output from the instrument

Data from a ground-based gamma radiometric survey (Figure 3.9) shows different spatial patterns in each individual layer. A comparison with the same season yield map (Figure 3.7) points to the potassium (K) and the total count (TC) layers (c and d) as most closely representing final yield. The similarity between the two layers is due to the gamma emissions from the breakdown of potassium-based minerals and potassium linked to clays dominating the total emissions from the field.

Using gamma radiometrics in PA

Gamma radiometers can be deployed proximally or from aircraft. When deployed proximally at ≤40 m swaths, the potential exists to measure short-range gamma-ray emission variation in fields and farms that is indicative of changes in soil-forming processes and soil parent material. This information can be a key component for producers wishing to manage within-field variation.

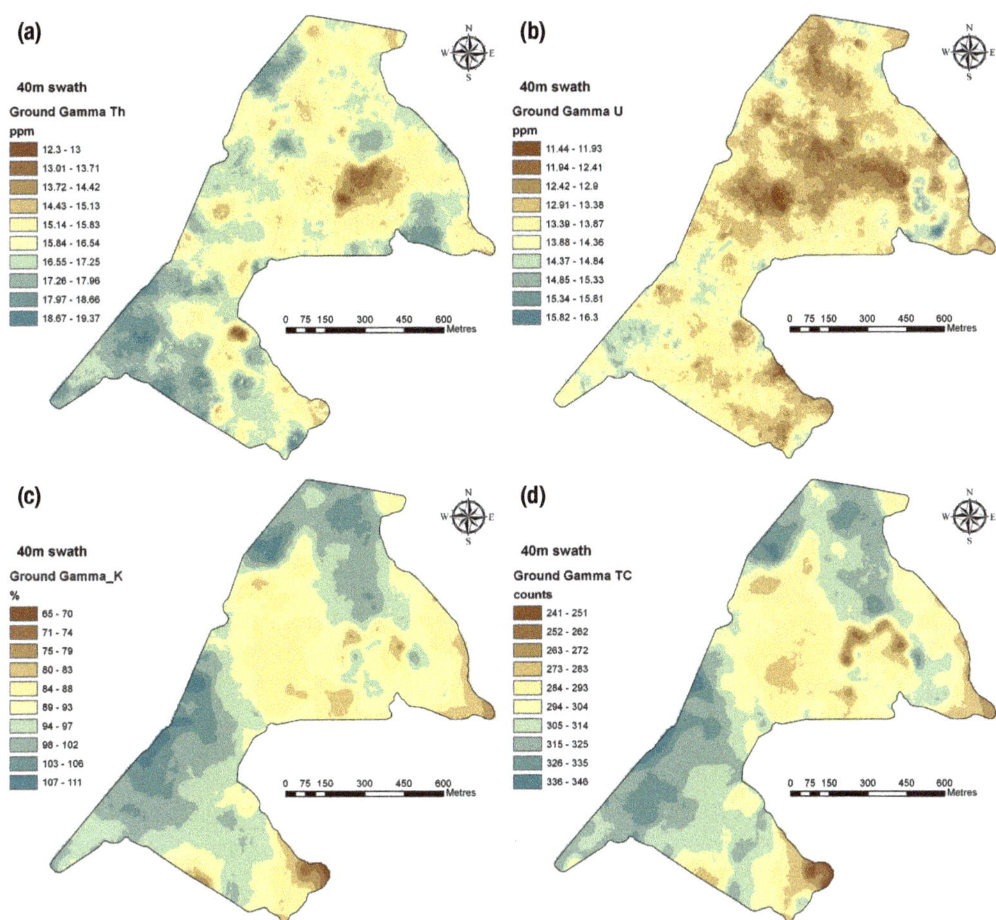

Figure 3.9: Data gathered from a ground-based gamma radiometric survey. (a) Thorium. (b) Uranium. (c) Potassium. (d) Total gamma count.

Useful relationships have been found between gamma radiometrics and geomorphology, soil properties (particularly plant available potassium), topsoil texture, unsaturated hydraulic conductivity, bulk density, organic carbon, Colwell phosphorus and pH. As with EC_a measurements, these relationships are specific to the areas in which they were developed and cannot be used universally.

Proximal soil sensor development

While an EC_a or gamma emissions survey can be commercially obtained, a range of other sensing technologies continue to be explored for use in proximal, on-the-go soil property measurement. Examples of a soil pH sensor and a number of soil reflectance sensing systems have been released but their use in the provision of commercial soil surveys is minimal. Tables 3.3 and 3.4 respectively list the

Table 3.3: Currently available and potentially useful techniques for proximal, on-the-go monitoring of important soil chemical properties

Soil properties	Limitations to yield	Proximal sensing techniques that show potential	Conventional methods for calibration or ground-truthing
Soil nutrients	Deficiency, e.g. N, P, K, S and trace elements or toxicity, e.g. Al, B	Visible/NIR/SWIR/MIR spectroscopy Ion-selective electrodes Ion-selective field-effect transistor (ISFET)	Laboratory-based soil nutrient test Laboratory-based plant tissue nutrient test Crop visual indicators
Soil pH	Nutrient availability and Al and B toxicity	Ion-selective electrodes ISFET	Laboratory-based test for soil pH
Organic matter	Low organic matter	Visible/NIR/SWIR/MIR spectroscopy	Laboratory-based test for organic carbon Laboratory-based NIR/MIR spectroscopy
Soil sodicity	High sodium content	EMI Resistivity	Laboratory-based test for soil dispersion Laboratory-based test for CEC
Soil salinity	High salt content	EMI Resistivity Ground-penetrating radar	Laboratory-based test for electrical conductivity Crop visual indication of growth patchiness

currently available, and potentially useful, techniques for on-the-go monitoring of important soil chemical and physical properties.

Terrain sensing

All GNSS receivers are able to determine elevation when calculating location coordinates. So elevation data can usually be collected during other operations using a GNSS (e.g. harvesting or spraying). However, GNSS receivers have more difficulty determining the elevation component of a location, and as a general rule-of-thumb the vertical error is usually 1.5–2 times greater than the horizontal error. With this additional error, it is generally not possible to obtain good elevation data from stand-alone receivers. At a minimum, a differential correction is required to produce reasonable quality elevation data. For elevation mapping in flat landscapes, carrier-phase measurement is needed. Figure 3.10a shows an example of original elevation data collected using a WADGPS (±10 cm 2RMS). Figure 3.10b shows a map produced from the data.

Information on the elevation or topography of fields is often useful in understanding production response. Topography is known to influence soil formation, water movement and cropping aspect in the landscape. To understand this impact, the elevation data gathered from a GNSS can be used to construct a digital elevation model (DEM) of a field or farm. This can then be used to identify specific terrain attributes such as slope, aspect, curvature, solar radiation

Table 3.4: Currently available and potentially useful techniques for proximal, on-the-go monitoring of important soil physical properties

Soil properties	Limitations to yield	Proximal sensing techniques that show potential	Conventional methods for calibration or ground-truthing
Soil texture/soil type	Low inherent yield potential due to low CEC, PAWC, inherent fertility	Gamma radiometrics EMI Resistivity Visible/NIR/MIR spectroscopy Ground-penetrating radar Tillage draft	Hand texturing of soil sample Laboratory-based particle size analysis (PSA)
Soil water storage capacity (PAWC)	Low water content	EMI Resistivity Visible/NIR/MIR/thermal IR spectroscopy Radar	Drained upper limit (DUL) estimates Crop lower limit (CLL) estimates
Soil water in season (PAW)	Low PAW	Thermal infrared Visible/NIR/MIR spectroscopy EMI Resistivity Radar Time-differential imagery	Laboratory-based mass balance measurements *In situ* neutron/capacitance/TDR moisture probes Estimate from soil texture
Waterlogging	Reduced oxygen availability	Elevation EMI Resistivity	Piezometers/dip wells Visual observation of crop chlorosis Surface water ponding Soil hydraulic properties
Rooting depth	Shallow rooting depth Abrupt changes to soil texture Subsurface compaction Rocks	EMI Resistivity Ground-penetrating radar	Soil pit profile description Manual push probe

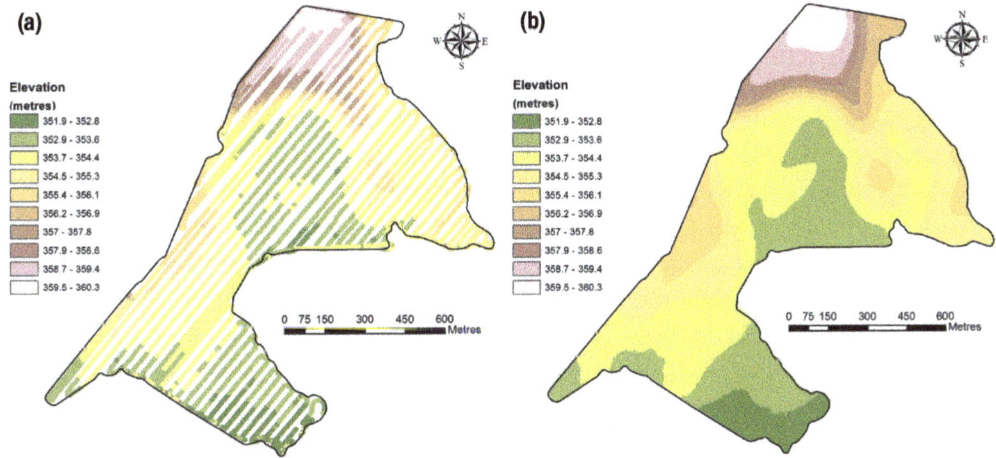

Figure 3.10: (a) Original elevation data captured using a WADGPS. (b) A map produced from the data.

interception, landscape water flow directions and topographic wetness indices. This analysis requires specialised software and analysis but it may provide important insights into crop behaviour. Often, though, a basic elevation map is sufficient to help identify the agronomic effects of topography variation on production variability or to create cut-and-fill maps for land levelling procedures.

Airborne and satellite optical imagery

Remote sensing is the broad term for any system that uses planes or satellites from which to collect data. In agriculture, it applies to gathering information on aspects of the landscape below. Mostly, remote sensing in PA uses optical imagery to record the spatial variability in soil and crops. During the growing season, vegetative growth may be monitored for production variability resulting from nutrient deficiencies, water stress or pest infestation, all of which may influence final yield. Imagery is also used to map farm boundaries, watercourses and terrain.

Optical imagery relies on the fact that areas with different soil and vegetation cover have distinguishing reflectance signatures in the visible and/or non-visible electromagnetic (EM) spectrum. The amount of energy reflected, transmitted or emitted in these areas of the EM spectrum provides information linked to many vegetation and soil characteristics. Optical remote sensing can provide a useful, low-cost means of assessing variability in these characteristics at a field, property and regional scale within and between seasons. In some situations these images can provide a surrogate production map.

Images can be captured using photographic film, video or digital media. Both aerial and satellite-based imagery have four properties which are important to their use in PA:

- spatial resolution – essentially the size of the smallest object that can be identified in an image. It is loosely defined by the picture element (pixel) size of the image;
- spectral resolution – the number of segments (spectral bands) of the EM spectrum that can be measured;
- radiometric resolution – the number of data levels for each band that can be stored;
- temporal resolution – the minimum time period between two images of the same area.

The spatial resolution of images captured by aerial platforms is generally a function of observation altitude and the capturing media, but is usually in the range of 1–2 m^2. Images captured from a satellite have a spatial resolution of 1–100 m^2, depending on the sensor. The spatial resolution used for agriculture has traditionally been in the range of 10–30 m^2. However, newer sensors and image-processing techniques have seen images of 1–2 m^2 resolution increasingly used.

Remote optical sensing systems involve passive optical sensors that use natural light from the sun to illuminate a surface, then record the reflectance of that light in a number of wavelengths. These wavelengths can range from visible light (wavelengths of 0.4–0.7 μm) through to the near-infrared (NIR, 0.7–1.3 μm) and shortwave (mid) infrared (MIR, 1.3–2.5 μm) to the thermal (long wave) infrared (LIR 8–14 μm) part of the EM spectrum.

Most optical systems are multispectral, measuring reflectance in three to 10 areas (spectral bands) of the EM spectrum. Satellite systems tend to have wide spectral bands, whereas airborne systems use narrow spectral bands. Many multispectral optical systems also have a single high spatial resolution panchromatic (grey-scale) band, which can be used to enhance the resolution of a multispectral image. This process is known as pan-sharpening. The most advanced optical sensors are hyperspectral, with 126–224 very narrow spectral bands, often less than 0.01 μm in width.

The radiometric resolution of these systems is commonly expressed as the number of bits (binary digits) needed to store the maximum data level. An eight-bit data set will be able to store 2^8 data levels (256 data levels), an 11 bit data set will store up to 2^{11} (2048) data levels and a 16 bit data set will store up to 2^{16} (65 536) data levels. In essence, the higher the number of data bits, the more subtle the changes that can be seen in an image.

Optical sensing is affected by atmospheric conditions such as cloud, thick haze, dust or smoke. Thermal and shortwave infrared bands are less susceptible to haze, smoke or dust than are the visible and NIR bands.

Optical airborne imagery

Gathering digital imagery using aircraft platforms provides flexible timing of acquisition, which can be very important in cropping systems where critical crop development stages and management operations are controlled by seasonal conditions. The data is gathered at high spatial resolution (25 cm–3 m), with narrow spectral bandwidths and high radiometric resolution (commonly 10–16 bit). The spatial resolution is determined by the type of sensor and the elevation of the aircraft. Lowering the survey altitude results in a higher spatial resolution but it reduces the ground area coverage (scene size) of each image. The data is less affected by general atmospheric conditions than is satellite imagery, because the sensors are operating much closer to the Earth's surface.

Airborne digital imagery does have some disadvantages, when compared to satellite acquisition. Covering large areas can be difficult due to the relatively small scene size and the potential need to merge neighbouring scenes. Merging scenes may be an issue because an aircraft is a less stable platform than a satellite, and variation in altitude and attitude (roll, pitch and yaw) can produce spatial distortion and brightness differences between each captured scene. Most major cropping areas in Australia can be covered by commercial aerial imagery services, however, costs vary depending on the distance between target area and aircraft base.

As a rule-of-thumb, aerial imagery is usually captured one and a half hours either side of solar noon to provide sufficient illumination and minimise shadowing. And, while an aircraft may be able to operate below high cloud cover, any cloud will reduce the optimal window for image capture. In winter, lower solar angles further restrict the timing of flights.

Most airborne multispectral systems use a sensor that records three to four spectral bands, although some have up to 10. The use of hyperspectral sensors is less common because the sensors are more expensive and analysis of the imagery needs to be far more detailed to extract benefits. However, the higher spectral resolution of these images offers opportunities to discriminate plant characteristics and functional properties, and surface soil characteristics, that multispectral sensors cannot offer.

Optical satellite imagery

The increasing array of satellite platforms which provide optical imagery (Table 3.5) ensures that optical data can be gathered anywhere on the planet. Together with improvements in sensor technology, this range means that measuring crop and soil reflectance characteristics over moderate to large areas is easily achievable. The spatial resolution (pixel size) varies considerably between systems:

- high-resolution imagery (e.g. IKONOS, Quickbird, RapidEye) may be 1–5 m;
- moderate resolution imagery (e.g. SPOT, Landsat) is 10–30 m;
- very coarse imagery (e.g. MODIS) is between 100 m and 1 km.

Table 3.5: Basic specifications for satellite-based remote sensing systems of potential use in PA

	Spectral resolution (bands)	Spatial resolution (m)	Temporal resolution (days)
MODIS	36 bands R, NIR B, G, MIR (0.4–14.4 μm)	 250 500 1000	1–2
ASTER	14 bands G, R, NIR 6 MIR bands 5 LIR bands	 15 30 90	On request
Landsat 5 TM	B, G, R, NIR, MIR LIR	30 120	16
Landsat 7 ETM+	Panchromatic B, G, R, NIR, MIR LIR	15 30 60	16
SPOT 4	Panchromatic G, R, NIR, MIR	10 20	1–4

Table 3.5: Continued

	Spectral resolution (bands)	Spatial resolution (m)	Temporal resolution (days)
SPOT 5	Panchromatic G, R, NIR MIR	2.5 10 20	1–4
Worldview-1	Panchromatic	0.6	1–2
Worldview-2	Panchromatic B, G, R, NIR Red edge Coastal Yellow NIR2	0.5 1.8 1.8 1.8 1.8 1.8	1
Resourcesat-2	G, R, NIR G, R, NIR, SWIR	5.8 23	5 24
RapidEye	B, G, R Red edge 3 NIR bands	5 5 5	1
IKONOS	Panchromatic B, G, R, NIR Pan-sharpened 4 band	0.8 3.2 1.0	3
QuickBird	Panchromatic B, G, R, NIR Pan-sharpened 4 band	0.6 2.4 0.6	1–3
Geoeye-1	Panchromatic B, G, R, NIR	0.4 1.7	2
EO-1	Hyperspectral 196 bands (0.36–2.78 μm)	30	16
Due for operation 2013 onwards			
Pleiades–HR1 & 2 (HR1 launched December 2011; HR2 launch 2013)	Panchromatic B, G, R, NIR	0.5 0.5	1
Landsat Data Continuity Mission (LDCM – Landsat 8) (launch 2013)	Panchromatic B, G, R, NIR, MIR LIR	15 30 60	16
Geoeye-2 (launch 2013)	Panchromatic B, G, R, NIR	0.3 1.7	1

The radiometric resolution of the imagery also varies between systems. The older Landsat and SPOT systems have eight-bit sensors, while imagery from some of the newer satellite systems (e.g. IKONOS, Quickbird, RapidEye) is 11 bit.

A system with a higher radiometric resolution is able to store more levels of data which, aside from assisting in recording more subtle changes in reflectance, means the sensor can still distinguish differences at higher levels of reflectance. The practical importance of this is that plant reflectance can be measured further into the growth cycle before the image is saturated by strong reflectance. A higher radiometric resolution, as with higher spatial resolution (small pixel size) or increased spectral resolution (more bands of the spectrum), results in larger storage requirements for the data.

One of the major historical limitations to the use of satellite imagery in PA has been the temporal resolution of the older satellite systems (e.g. Landsat). Long revisit times reduce the opportunities to monitor reflectance from crop and soil; if cloud cover interferes, the observation windows are further decreased. Many of the newer satellite systems have sought to reduce the revisit time. This has been achieved by engaging multisatellite systems and/or systems with sensors that can be pointed to the side of their standard flight path to gather data to a specific order.

A significant advantage of the Landsat system is the extent of its data archive. The Landsat program has been operational since the mid 1970s, with 30 m resolution raw path-oriented data scenes of 170 × 183 km available from the mid 1980s. Landsat imagery is often processed to a map-oriented 25 m resolution and sold as smaller sections of data for farm-scale use. This provides a valuable resource for PA that can be used, among other things, to benchmark and model changes in production across time.

Replacement missions, launching new satellite technologies, are ongoing (Table 3.5). This ensures that scene costs will continue to decrease while spatial, radiometric and temporal resolutions are being enhanced. These missions include systems and sensors that will prove useful in PA.

Using optical remotely sensed imagery

The selection of an imagery system should depend on how well the imagery specifications meet actual requirements for each specific use. A checklist of imagery specifications for reference when choosing suitable systems would include:

- spatial resolution;
- availability and timing of delivery;
- radiometric resolution;
- area captured in the imagery;
- temporal resolution;
- price of imagery and any required processing.

A comparison of the difference in detail available from optical sensors with different spatial resolutions is shown in Figure 3.11. This highlights the primary importance of obtaining the correct spatial resolution for a task.

Analysing digital imagery

There are many ways of analysing digital imagery that provide more information than just looking at them as a colour or grey-scale 'photograph'. These analyses

Figure 3.11: Comparison of the effects of different spatial resolutions on the detail observed in a wheat crop. (a) 25 m resolution. (b) 1 m resolution.

make use of the different bands of information captured by the sensors. Using these bands to create ratios and indices is an uncomplicated process that provides a very effective analytical tool. In PA, the most common bands used are the red and NIR bands because they are specifically, and sensitively, related to plant physiology.

In healthy, actively growing plants, blue light (0.4–0.5 µm) and red light (0.6–0.7 µm) are strongly absorbed by the chlorophyll pigment inside leaf cell chloroplasts to provide energy for photosynthesis. NIR light (0.7–1.3 µm) is strongly reflected in healthy, actively growing plants due to a robust and expansive internal leaf cell structure. Green light (0.5–0.6 µm) is predominantly reflected by all plants, so a healthy, vigorous plant appears green because of the greatly reduced amounts of red and blue light being reflected.

Healthy, actively growing plants will therefore display a large disparity between low red light and a high NIR light reflectance. In contrast, plants under stress typically absorb much less red light as the chlorophyll reduces in activity (red reflectance rises); as the leaf internal structure collapses, NIR reflectance decreases. As a result, the disparity between the red and NIR reflectance falls significantly.

Bare soil can be distinguished from healthy, actively growing plants because it typically has a higher red and lower NIR reflectance.

While it may be possible for the human eye to discern areas of more actively growing crop as 'more green' within a true colour image of a crop field (Figure 3.12a), using the red and NIR band differences proves more discriminating.

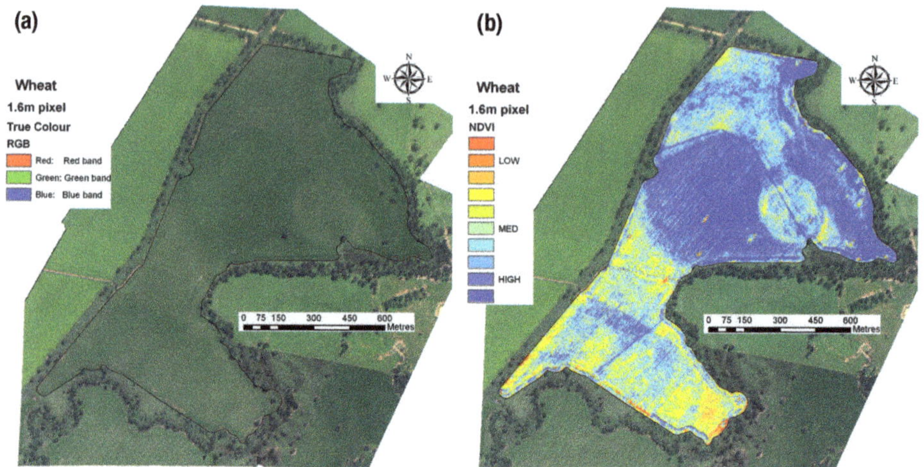

Figure 3.12: (a) A true colour image made by combining the red, green and blue bands shows areas of greater vigour as greener. (b) An NDVI map that accentuates the differences in crop performance between different areas of the field.

A range of ratios has been developed using the reflectance in these bands, the most common of which is the Normalised Difference Vegetation Index (NDVI). This aims in its calculation (Table 3.6) to further exploit the disparity in red and NIR reflectance. This ratio process creates a single band image where the higher the value in each pixel, the greater the vigour, health or greenness of the crop (Figure 3.12b).

Vegetation indices have been developed for other combinations of sensor band data gathered by aerial and satellite platforms. Some of the most common are included in Table 3.6 along with the inferred physiological relationships. Generally, indices created using narrow spectral bands can provide more physiologically specific information. As sensor technology and crop reflectance research improves, new indices will continue to be proposed. A recent example is the development of the red edge NDVI, which is more commonly referred to as NDRE (Table 3.6).

Proximal crop reflectance sensors

These sensors are generally mounted on ground-based vehicles and in essence work in the same fashion as the sensors that are used to gather aerial or satellite imagery. The main difference is that the majority of new-generation versions use their own light source to illuminate the crop, and are therefore 'active' systems. This means they can be used 24 hours a day, as they are not dependent on the sun as a light source. These sensors build an image of the whole field as they traverse.

Table 3.6: Common vegetation indices used in remote crop sensing. NDRE is calculated from narrow band data with wavelengths indicated in subscripts

Index	Bands used	Physiological interpretations
Vegetation (Simple) Index (VI)	$\frac{IR}{Red}$	Crop greenness, vigour, leaf area
Normalised Difference Vegetation Index (NDVI)	$\frac{iR - Red}{IR + Red}$	Crop greenness, vigour, leaf area
Soil Adjusted Vegetation Index (SAVI(0.5)) (0.5 = correction for soil reflectance)	$\left(\frac{IR - Red}{IR + Red + 0.5}\right) \times (1 + 0.5)$	Crop greenness, vigour, leaf area when ground cover is sparse
Enhanced Vegetation Index (eNDVI)	$2.5 \times \frac{IR - Red}{IR + 6 \times Red - 7.5 \times Blue + 1}$	Crop greenness, vigour, leaf area in crops with high reflectance
Photosynthetic Vigour Ratio (PVR)	$\frac{green}{red}$	Strong chlorophyll absorption (photosynthetic activity)
Plant Pigment Ratio (PPR)	$\frac{green}{blue}$	Strongly pigmented crops
Plant Cell Density Ratio (PCD)	$\frac{IR}{green}$	Density of healthy plants cells
Green Normalised Difference Vegetation Index (GNDVI)	$\frac{IR - green}{IR + green}$	Chlorophyll content, cell density and stress
Red edge NDVI (NDRE)	$\frac{IR_{750} - Red_{705}}{IR_{750} - Red_{705}}$	Crop greenness, chlorophyll content, water stress

The information gathered by the sensor is immediately converted to a reflectance index. This can be tagged with location information from a GNSS and stored to make a map for later use, or used in conjunction with variable-rate technology (VRT) to control changes in fertiliser or chemical application while driving across the field.

Figure 3.13 provides an example of NDVI data obtained from a digital aerial image compared to that gathered by a proximal crop reflectance sensor system.

There are several proximal crop reflectance sensors commercially available in Australia as at 2013. A general overview of their specifications is provided here, but readers are advised to contact the manufacturers or local distributors to check for any modifications.

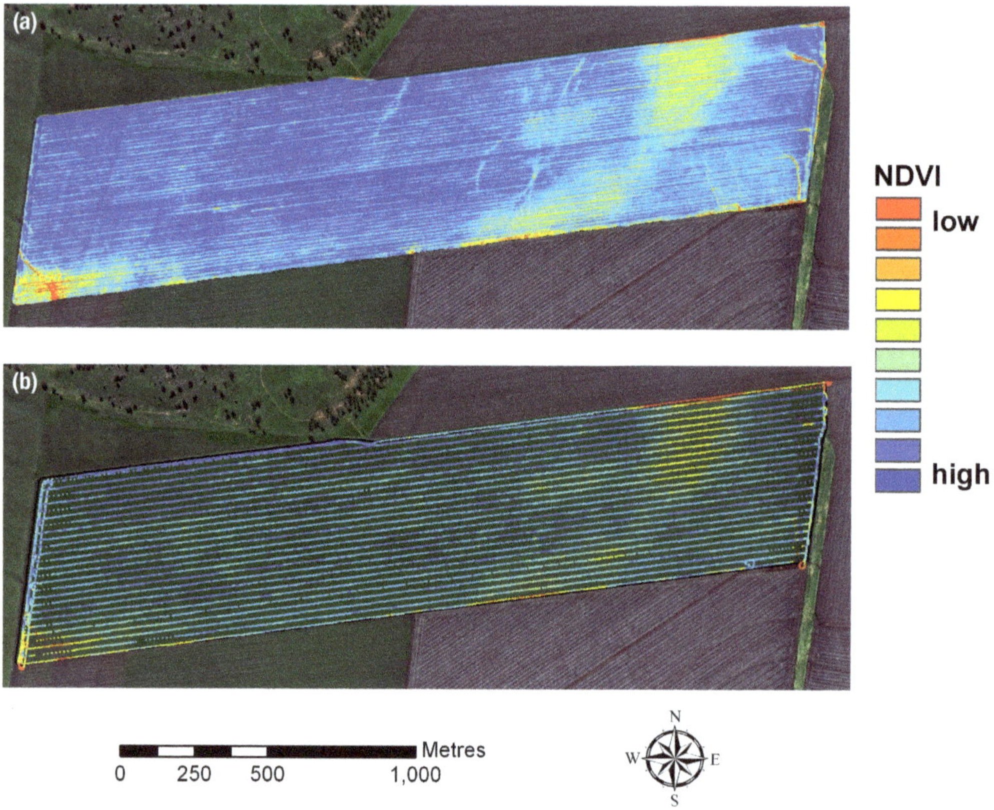
(a)
(b)
NDVI
low
high
Metres
0
250
500
1,000
N
W
E
S

Figure 3.13: A comparison of NDVI calculated from (a) an aerial image with 2 m spatial resolution and (b) reflectance data obtained using proximal crop reflectance sensors on a 27 m swath.

Crop Circle™

Model ACS-430
Crop Circle

Crop Circle

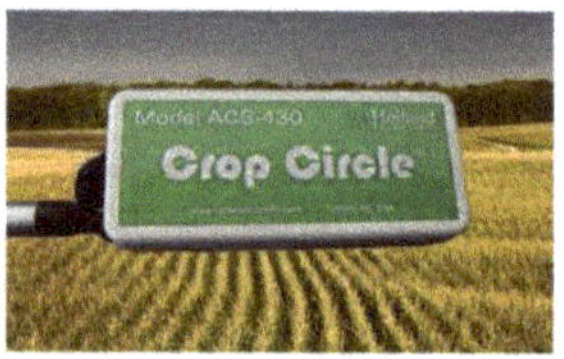

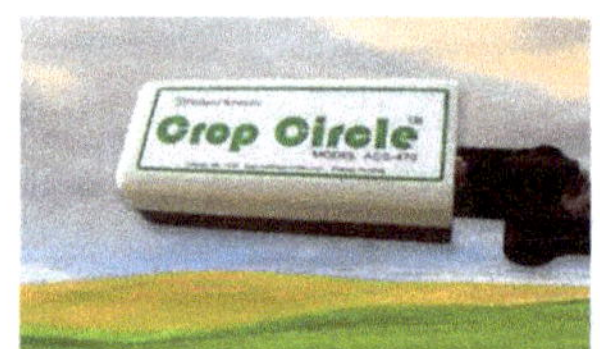

Company	**Holland Scientific – *(images courtesy of Holland Scientific)***
Height of operation	**0.25 m to 2.5 m**
Field of view	**Height x 0.6 (up to 8 sensors on CANbus)**
View angle	**nadir**
Active light source	**Model ACS430: Red (670 nm) or Red edge (730 nm) & NIR (780 nm)** **Model ACS-470: 3 user-configurable bands (430–800 nm)**
Data output	**Model ACS-430: band information and NDVI or NDRE** **Model 470: band information and user-defined index**
Calibrations	**Crop biomass and nitrogen uptake**

OptRx™

Company	Ag Leader
Height of operation	0.25 m to 2.5 m
Field of view	Height x 0.6 (up to 8 sensors on CANbus)
View angle	nadir
Active light source	Red (670 nm) or Red edge (730 nm) & NIR (780 nm)
Data output	Vegetation Index (NDRE or NDVI) and nitrogen recommendations
Calibrations	Corn, wheat and user defined for other agrochemicals

GreenSeeker®

Company	Trimble Navigation Limited – *(image courtesy of Trimble)*
Field of view	0.6 m (multiple sensor capable)
Height of operation	0.6–1.6 m above target
View angle	nadir
Active light source	Red (656 nm) & NIR (774 nm)
Data output	NDVI or four alternatives and nitrogen recommendation
Calibrations	Yield potential and nitrogen responsiveness – winter wheat, spring wheat, canola, corn, sorghum, cotton

N-Sensor® (ALS)

Company	Yara International ASA – *(image courtesy of Yara)*
Height of operation	Tractor cab height
Field of view	3 m wide strip on each side of the tractor
View angle	oblique
Active light source	Red edge (730 nm) and (760 nm)
Data output	Biomass index and nitrogen recommendation
Calibrations	winter wheat, winter barley, spring wheat, spring barley, canola, winter rye, corn, triticale, oats, potato, protein in winter wheat

CropSpec™

Manufacturer	TOPCON Precision Agriculture
Height of operation	Tractor cab height
Field of view	3 m wide strip on each side of the tractor
View angle	oblique
Active light source	730–740 nm and 800–810 nm
Data output	Biomass Index and N recommendation
Calibrations	winter wheat, winter barley, spring wheat, spring barley, potatoes, protein in winter wheat

WeedSeeker®

Company	Trimble Navigation Limited – *(image courtesy of Trimble)*
Field of view	Model 650: 0.3 m (multiple sensor capable) Model 655: 0.38 m (multiple sensor capable)
Height of operation	Model 650: 0.45–0.75 m; Model 655: 0.61–0.75 m
View angle	Nadir
Active light source	Red (660 nm) and NIR (770 nm)
Data output	NDVI
Calibrations	Green plants on soil or stubble

WEEDit®

Manufacturer	Rometron B.V. (*images courtesy of Rometron BV)*
Height of operation	1 m
Field of view	100 cm, divided in 5 sections of 20 cm
View angle	nadir
Active light source	Red LED
Data output	Detects fluorescence from chlorophyll
Calibrations	Green plants on soil or stubble

Table 3.7: The main types of remote and proximal reflectance sensing techniques available for PA, and their applications

		Attribute estimated		
Property observed	**Platform**	**Soil**	**Crop**	**Other**
Visible/NIR/ shortwave infrared (SWIR) reflectance	Proximal Aircraft Satellite	Moisture content Colour Organic matter Surface soil mineralogy (hyperspectral only)	Colour and vigour Leaf area index Biomass N status and potential responsiveness Photosynthetic activity Yield potential Crop type Stress Physical damage Insect and disease incidence Moisture content and senescence Harvest index or stubble quantity	Soil and plant tissue sampling locations Management classes Land use
Thermal infrared reflectance (TIR)	Proximal Aircraft Satellite	Moisture content	Canopy temperature Moisture stress Disease incidence Vigour Evapotranspiration	Soil and plant sampling locations
Radar	Aircraft Satellite	Moisture content Surface roughness Salinity Texture	Leaf area index Biomass Moisture content Crop type Crop structure and height Stubble quantity	Management classes Land use

Summary of reflectance sensors

Many reflectance properties can be measured remotely or proximally. Table 3.7 shows the main types of reflectance measurements and the attributes they are attempting to estimate. It is important to note that these tools need 'ground truthing' to calibrate them to the actual attributes of interest.

Variable-rate technology

Variable-rate technology (VRT) allows fertiliser, agrochemicals, lime, gypsum, irrigation water and other field-based mechanical operations to be applied at different rates across a field, without manually changing rate settings on equipment or having to make multiple passes over an area.

Such variable-rate application (VRA) means that changes in crop yield potential across a field can be matched with a specific input rate. Matching input to

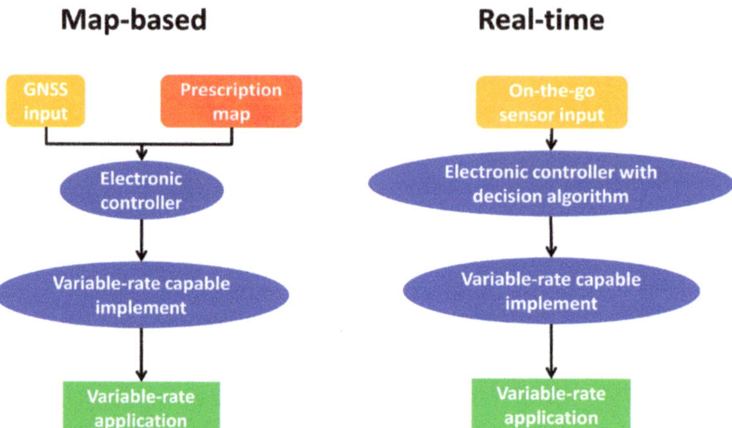

Figure 3.14: Schematic description of map-based and real-time VRT systems.

potential output should make the cropping operation more cost-effective and, by restricting inputs such as fertiliser and chemical to the crop requirement, should minimise any potential negative environmental impacts.

VRA operations can range from the simple control of flow rate to more complex management of rate, chemical mix and application pattern.

Two general systems for the use of VRT are schematically described in Figure 3.14.

Map-based control

A map of desired application rates or actions (prescription map) is produced for the field and loaded into the tractor prior to the actual operation. This requires a GNSS

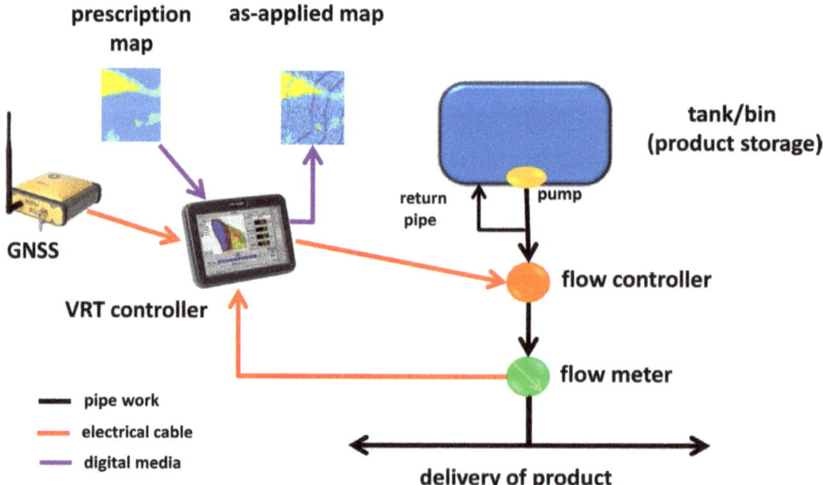

Figure 3.15: Schematic description of the components required for a generic map-based VRA system.

and a computer in the tractor to read where the vehicle is on the map and to communicate the required rate at a location to the rate-control unit. The rate-control unit then sends an appropriate signal to the actuator/motor/pump on the application implement, to perform the job at the required rate (Figure 3.15). Prescription maps can be produced for fertiliser, lime, gypsum or agrochemical application. The computer and the rate-controller can be integrated units or separate components.

Real-time control

The decision about what rates to apply where is made on-the-go, using information gathered during the actual operation. This requires a number of suitable sensors to detect the required information and pass it to a computer in the vehicle. The computer carries a calibration equation to turn the sensor information into a desired rate, which is then sent through a rate-controller to the actuator/motor/pump on the application implement. This type of system is usually designed for a specific job, such as herbicide or nitrogen fertiliser application. A GNSS is only required to keep a record of what rate was actually applied where – an 'as-applied' map.

Input auditing

Figure 3.16 shows an example of a prescription map (3.16a) and an as-applied map (3.16b) for a nitrogen fertiliser application. An as-applied map is constructed from information gathered in a feedback loop through the variable-rate controller and the actuators/motors/pumps involved in the application process. They can be made using map-based or real-time control. The data is gathered by the variable-rate controller. It allows managers to check what rates were actually applied. This is important for identifying any mistakes, that may be corrected if necessary or

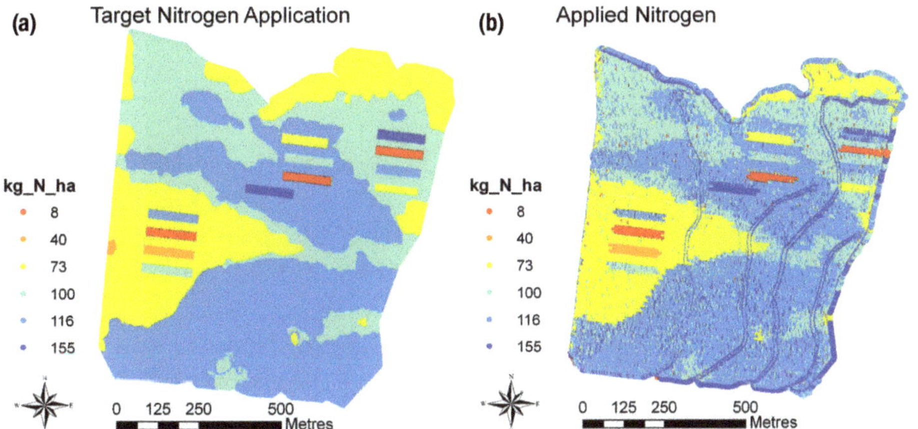

Figure 3.16: (a) A prescription map. (b) An as-applied map, both for a nitrogen response trial laid out using VRT. Contour banks are evident in the as-applied map.

Table 3.8: VRT operational controls for a range of management practices

Variable-rate management practice	Aspect	Technology
Fertiliser application	Spreading	Master control governing: • belt speed • feed gate height
	Pneumatic	Master control governing: • electric drive • electronic over hydraulic • electronic over mechanical
	Gas/liquid	• Master control governing flow controllers
Gypsum/lime application	Spreading	Master control governing: • belt speed • feed gate height
	Injection	• Master control governing flow controllers
Sowing	Quantity	Master control governing: • electric drive • electronic over hydraulic
	Depth	• Master control governing electronic over hydraulic
Spraying	Quantity and mix	• Master control governing electronic over mechanical
Irrigation	Sprinkler	Master control governing: • electronic over mechanical • electronic over pneumatic

feasible, and allows a record to be kept of the location to track production consequences. The record-keeping aspect is important for agrochemical application and may play a role in environmental auditing in the future.

Table 3.8 shows the commercially available technology options for VRT, based on different management operations.

Multi-sensor and autonomous platforms

The continued development of proximal and invasive sensing systems for on-the-go measurement of soil, crop and environmental variables is set to lead the technological development of SSCM. As more sensors become available, deploying them in unison offers improvements in data-gathering efficiency and costs. Multi-sensor deployment obviously depends on grouping sensors which can be physically and agronomically operated at the same time during the cropping season.

An added benefit may be the ability to combine the observations from a number of sensors into new diagnostic techniques for properties that are more

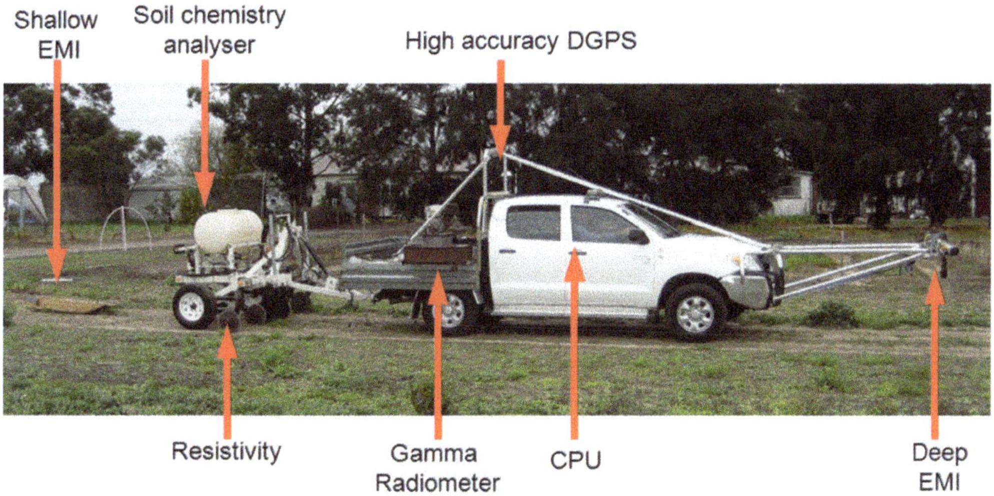

Shallow EMI | Soil chemistry analyser | High accuracy DGPS | Resistivity | Gamma Radiometer | CPU | Deep EMI

Figure 3.17: A multi-sensor system for on-the-go measurement of a number of soil and landscape properties in one operation.

difficult to directly measure, such as profile soil moisture, total crop nitrogen content and weed species identification. Potential sensor groupings include the pre-sowing use of GNSS elevation, EC_a, gamma emissions, soil nutrients, soil strength and reflectance-based weed spraying, and the in-season use of GNSS elevation, EC_a, gamma emissions, crop reflectance, crop fluorescence and radar.

An initial, if crude, attempt at a multi-sensor system for measuring soil properties and GNSS elevation on-farm is shown in Figure 3.17. A more elegant example can be found in the Mars exploration rover *Opportunity*, deployed by NASA, which has been sampling and analysing soil and environmental variables autonomously for over eight years. Its 2012 replacement, *Curiosity*, is equipped with multi-spectral digital imaging cameras, four different spectrometers (including a laser-induced breakdown spectrometer (LIBS)) and a gas chromatograph for elemental analysis of samples, a neutron-based instrument for measuring ground moisture, an environmental (weather and radiation) monitoring station and stereoscopic 3D imaging cameras for sample collection, navigation and hazard avoidance.

It is not difficult to envisage that the instruments and designs of such exploration vehicles will have an impact on agriculture. They combine technology for sensing systems and controllable operations, which could be employed in mapping and variable-rate applications in SSCM. It should lead to the combination of soil, crop and environmental sensors on platforms dedicated for farm use which will become more unobtrusive and will eventually monitor and, where appropriate, treat autonomously.

Key points

Yield monitors

- A grain yield monitoring system includes a grain flow sensor, grain elevator speed sensor, grain moisture sensor, harvester speed sensor and a comb up/down sensor. All are connected to a monitor and a global navigation satellite system (GNSS) if data is to be mapped.
- Yield can be calculated each second along the harvester path.
- To be most effective, yield monitors need to be calibrated properly. Calibration should be performed for each type of grain at the beginning of the harvest season. Calibrations should be checked during harvest.
- The accuracy of yield mapping is affected by various factors including incorrect installation of sensors, incorrect yield monitor calibration, incorrect harvester set-up and poor harvesting techniques.

Quality monitors

- Crop quality may be as important as crop yield in determining profitability, particularly for higher-value specialty crops such as durum wheat and malting barley.
- On-harvester grain quality sensors are relatively new and use near-infrared spectroscopy (NIRS) to estimate the protein, oil and moisture content of the grain.
- Maps of grain quality can be used to identify areas where nutrient or soil management may need to be changed or to provide information to segregate grain to meet specific market requirements.

Soil and terrain sensing technology

- Apparent soil electrical conductivity (EC_a) is the predominant soil property measured by on-the-go systems. This is done using electromagnetic induction sensors or electrical resistivity sensors.
- EC_a is affected by the soil texture (i.e. the amount of clay), the type of clay, the soil moisture content and the amount of nutrients or salts dissolved in the soil water.
- Generally, the soil EC_a can be used to estimate yield potential, subsoil constraints or salinity.
- High-resolution GNSS (e.g. RTK-GPS) can provide an accurate map of elevation across a field, which gives information about how water moves across the landscape, together with changes in the aspect of the crop relative to the path of the sun.

Remote sensing

- Uses sensors mounted on aeroplanes or satellites to gather information about soil and crops.
- Its most common use in PA is to collect an image of a crop to assess the amount and quality of plant growth.
- The detail in the image depends on the altitude and type of sensor being used.

Crop and weed reflectance sensors

- Crop reflectance sensors are generally used to estimate the crop biomass and vigour during the growing season. Such sensors are being used to control the application of nitrogen fertilisers.
- The same technology can be used to detect the presence and location of green weeds in fallow fields and to target herbicide sprays. Shielded spraying nozzles can extend this operation to green weeds between crop rows.

Variable-rate technology

- Variable-rate technology (VRT) provides a way of changing input rates as vehicles move across a field. Inputs that may be variably applied include fertiliser, lime, gypsum, agrochemicals, seed and irrigation water.
- Variable-rate application of inputs can be controlled from predefined prescription maps or from real-time crop or soil sensors.
- Variable-rate systems require:
  - a prescription map or on-board sensor system to provide site-specific input rates;
  - a GNSS to relate the applicator to the prescription map and/or to map what was actually applied;
  - a variable gearbox, actuator, servo valve or injection pump that controls the output quantity of the application equipment;
  - a variable-rate controller that drives the gearbox/actuator/valve/pump;
  - a computer that supplies the correct rates to the controller (in some systems, these last two are combined).
- Calibration of application implements is still essential.

4

Software for Precision Agriculture

This chapter provides an introduction to the issues that should be considered when purchasing software to perform PA tasks. It deals with generic issues and documents the diversity in relevant software. However, given the rapid evolution of computer software, only brief information is provided on capabilities, with the aim of helping to identify potentially suitable software packages for further investigation. It is strongly recommended that an up-to-date search be conducted when purchasing software. The options discussed here are given as a guide and are as complete as possible, but no endorsement is given to any particular program and nor is there censure of any software not mentioned. The allocation of software to categories is intended as a guide only. Each particular enterprise requires a software program that best suits its needs and budget.

Software for PA

Software is currently available to perform the following PA tasks:

- farm record-keeping and accounting;
- global navigation satellite system (GNSS) mapping or vehicle navigation;
- yield data processing, mapping and storage;
- image data processing and storage;
- soil and other spatial data processing, mapping and storage;
- spatial data analysis and management decision support;
- prescription map building and export;
- management and instruction of variable-rate control devices;
- analysis of harvest trial data.

Some programs are developed specifically for a single purpose, often as a stand-alone piece of software. Other programs are capable of performing multiple tasks.

Before purchasing software a grower or consultant should list all the features/tasks that they require now, and consider future potential requirements.

This type of wish-list will identify the most suitable program(s). In many cases, a number of programs or add-ons/modules will be required to meet all the needs of an end-user.

Proprietary file formats

The different brands of navigation, monitoring and controlling systems being used in PA are commonly operated by software that is unique to the brand. In the vast majority of cases this means that the file types needed to instruct the equipment and to record information are also unique to the brand. These proprietary files may cause compatibility problems between software and hardware and between software programs. Obviously, proprietary software should work with same-brand equipment. Users should enquire about compatibility with existing hardware and software when seeking to use third-party task-based software or all-in-one data management software.

Task-based software

These software tools are designed to complete a single task or a small number of specific tasks. They are usually developed to achieve one or more of the following functions:

- perform an operation on a specific piece of hardware;
- complete basic operations or view maps without comprehensive software;
- perform an operation that is missing from the more comprehensive software programs;
- be part of a modular software suite that can be expanded as a user's requirements change or increase;
- operate on mobile computer platforms.

The software for operation of specific hardware is usually supplied by the manufacturer of the hardware (Table 4.1) and is included in the purchase price. Some hardware manufacturers have developed a modular style of software application; the cost may be applied as each module is acquired. There are a number of freeware or shareware software options available for PA applications (Table 4.2).

Table 4.1: Examples of task-based software developed for PA implement/sensor control and data management

Company	Software	Description
AGCO	Falcon II Application Control	VRA control
AgLeader	Direct Command	VRA control
	Seed Command	VRA planter and seeding control
Case IH	AFS Planter Software	VRA planter control
	AFS Pro600 Advanced	Yield data recording and mapping
Farmscan	Farmlap2e	Steering, VRA control and boom spray control
	Grain Yield	Yield data recording and mapping
	Irrigation Manager	VRA pivot control
Farm Works	Lexion Yield Tools	Yield data processing and mapping for Lexion
John Deere	Field Doc suite	Various VRA control
	Harvest Doc Combine	Yield data recording and mapping
Mapshots	EASi Rx	Converting Shape (.shp) to VRA files
Micro Trak	Data Trak	Yield data recording and mapping
Trimble (Greenseeker)	RT200 Software	VRA nitrogen control
Raven	Autoboom suite	Boomspray control
	SCS suite	VRA control
	Viper suite	VRA control
Trimble (Rawson)	Accu-rate software	VRA control
RDS	Apollo	VRA spreader and boomspray control
	Ceres 8000i Yield Module	Yield data recording and mapping
Topcon	X20 Maplink + VRC	Steering, VRA control and boomspray control

As the number of sensors supplying data for PA has increased, so have the software requirements. Research into the best use of the data has led to the identification of new analysis techniques, that have often required the development of task-based software to prove their usefulness (or otherwise). Including these analytical techniques in the more comprehensive software takes time, so there will always be a range of task-based software tools that can be implemented by more experimental users.

Other users may wish to build an analytical process on the basis of a number of software programs, some of which have not been designed specifically for PA (e.g. GEOD, FuzMe). These task-based programs are used for generic manipulation of spatial information or to access a proprietary data file format. They allow PA data to be used in commercially available statistics packages or allow the output from expensive geographical information systems (GIS) to be converted to a format

Table 4.2: Examples of task-based PA freeware or shareware developed for data management or viewing

Company	Software	Description
Precision Agriculture Laboratory (PA Lab)	Vesper	Spatial prediction for mapping
	FuzMe	Clustering for management classes
Delta Data Systems	Viewpoint	Map viewing and basic interaction
ESRI	Arcexplorer	Map viewing and basic interaction of ESRI format files
Map Maker	Map Maker Gratis	Map viewing and editing
	FOViewer	Read and export data from proprietary yield and application log files
MicroImages	TNTmips Free	Map viewing and basic interaction of TNT format files
NSW Lands	GEOD	Coordinate conversion between projections and datums for Australia
Pitney Bowes	MapInfo ProViewer	Map viewing and basic interaction of MapInfo format files
TerraGo Technologies	TerraGo Desktop	GeoPDF map viewing
University of Nebraska-Lincoln	Yield Check	Process and clean yield data in text (.txt) files
USDA ARS	Management Zone Analyst	Clustering for management classes
	Yield Editor	Process and clean yield data in text (.txt) files
Zonum Solutions	Shape2text	Convert Shapefiles (.shp) to text files

usable in PA hardware. Working with software this way can be a little more challenging than using a dedicated PA software package.

Mobile applications

These are software applications designed for hand-held devices, e.g. tablet PCs or PDAs (Table 4.3) that are primarily used in the field. They are often developed in conjunction with a desktop application with which they are designed to communicate and transfer files. Operating different brands of desktop and mobile software may result in compatibility problems if proprietary format files are involved. When considering the use of this type of software in conjunction with a desktop application, always enquire about file exchange compatibility.

Data processing packages (data management, analysis, action options and archiving in a single program)

Levels of complexity

There is a wide variety of software programs and they vary in complexity, cost and compatibility. These software packages enable the management and analysis of

Table 4.3: Examples of task-based PA software developed specifically for mobile computer applications

Company	Software	Description
AgCode	Total Field	Scouting, recording, mapping and display
AgLeader	SMS Mobile	Scouting, recording, mapping and display
AgRenaissance	PocketRecon	Scouting, recording, mapping and display
Delta Data Systems	Pocket DLog Mobile DLog	Scouting, recording, mapping and display
	Pocket spreader	VRA control
ESRI	ArcPad	Scouting, recording, mapping and display
Fairport	PocketPAM	Scouting and recording
Farm Works	Trac mate	Geo-referenced data entry
	Guide mate	Guidance
	Site mate basic	Farm mapping
	Site mate scouting	Scouting and farm mapping
	Site mate VRA	VRA control – one input
	Site mate multi-VRA	VRA control – multiple
GK Technology	Pocket Field Recs	Scouting and farm mapping
	Pocket VR	VRA control – multiple
SST	Summit and Stratus	Scouting, recording, mapping and display
StarPal	HGIS	Scouting, recording, mapping and display
Trimble	AgGPS EZ-Map	Scouting, recording, mapping and display

data from a variety of sources and the production of input-rate prescription maps for a variety of rate-controllers. There are three general levels of complexity to these software options for PA – introductory, intermediate and advanced – and the allocation to categories is offered as a guide only. The difference between the upper levels is blurring because, as more use is being made of PA in the farming community, the software from all companies is becoming more sophisticated.

Introductory level

Introductory-level software packages perform basic operations such as data retrieval and storage, simple mapping, basic data exploration and data export, but they do not provide detailed analytical tools (Table 4.4). They are primarily aimed at moving information from a data storage card or other media used in a farm vehicle, onto a computer. Many hardware manufacturers provide software tools at this level as an interface to their hardware, usually at minimal cost.

This introductory level of software suits growers who will outsource the majority of their data analysis to a consultant. Recent improvements have been the addition of database features that allow tabular summaries of harvests over fields, farms and years. Introductory-level packages may also allow the export of a machine-ready prescription file for use in the same brand or affiliated hardware.

Table 4.4: Examples of introductory-level single box software for data management, basic analysis, action and archiving

Company	Software	Description
AGCO	GTA200	Record keeping
	GA300	Record keeping and mapping
AGDATA	Phoenix Farms Mapping	Record keeping and mapping
BackPaddock	Manager	Record keeping, mapping and planning
CLASS	Agro Map Start	Basic yield mapping
	Agro Map Standard	Yield and farm mapping
eAgribusiness	iFarm	Record keeping and mapping
Fairport	gpMapper	Farm mapping
Farmscan	Data Manager	Yield mapping and data management
Farm Works	Site	Data management, yield and farm mapping
Leica	iNEX	Monitoring and farm mapping, recording and boomspray control
Rinex	PlanIT	Farm mapping, plans and data management
Rokit Science	Catchman	Record keeping, mapping and planning
SST	SSToolkit	Data management, yield and farm mapping
TeeJet	Fieldware suite	Data management, yield and farm mapping, VRA control
Trimble	EZ-Office	Data management, yield and farm mapping

Intermediate level

This level of software packages was originally provided by companies outside the major agricultural equipment manufacturers to meet the data analysis needs of growers and consultants in the PA industry. A number of equipment manufacturers now also sell branded software that fits this category, and the products are eminently suited to handling brand-specific PA data on the farm. The programs are generally user-friendly and farm-oriented with point-and-click features to perform common data analysis (Table 4.5). This software usually works with a wider range of file types, including various image file formats, than the introductory packages. Software in this category will generally perform simple operations with different data layers and can output user-defined variable-rate prescription maps. However, while they do not usually have the ability to build intricate decision support algorithms, they may allow for add-ons that can provide a specific functionality at an additional cost.

Advanced level

Advanced-level software programs have substantial display and analysis capabilities. The category includes software packages designed specifically for PA as well as more widely used generic GIS (Table 4.6).

Table 4.5: Examples of intermediate-level single box software for data management, analysis, action and archiving

Company	Software	Description
AGCO	GTA 400	Mapping, data management and analysis, VRA maps
AgLeader	SMS Basic	Mapping, data management and analysis, VRA maps
CaseIH	AFS Basic	Mapping, data management and analysis, VRA maps
CLASS	Agro Map Precision Farming	Mapping, data management and analysis, VRA maps
Fairport	FarmStar	Mapping, data management and analysis, VRA maps
Farm Works	Site Pro	Mapping, data management and analysis, VRA maps
GK Technology	Ag Data Viewer	Mapping, data management and analysis, VRA maps
LandView systems	LandView DSS Pro	Mapping, data management and analysis
Map Maker	Map Maker Pro	Mapping and data interrogation
MapShots	EASi Suite	Production records, data management and analysis, VRA maps
New Holland	PFS Advanced Mapping Software	Mapping, data management and analysis, VRA maps
SST	SSToolboxLite	Mapping, data management and analysis, VRA maps

Advanced PA/GIS programs often have a general user interface as well as an ability to write algorithms or subprograms for specific data analysis. These provide the greatest flexibility for working with a wide range of file types and analysis techniques. To make use of their capabilities, a higher level of computer literacy and commitment to learning their operational controls is required.

For many growers the advanced-level generic GIS programs may be unwarranted; however, for a consultant they can provide additional flexibility particularly for web-based applications. The dedicated PA software at this level is suitable for a farmer who is keen to gather spatial data on the farm from a number of sources and can spend time to understand, comprehensively analyse and build action plans with the data. Any agronomic consultants interested in PA should be aiming to use this type of software.

Decision support software

None of the data processing packages currently available for desktop computers include a comprehensive spatial decision support component. A full spatial decision support system for PA should be able to help with decisions on:

- identifying if and where there is manageable variability in production;
- determining causes of the variability;
- constructing and comparing potential management options.

Table 4.6: Examples of advanced-level single box software for data management, analysis, action and archiving

Company	Software	Description
Topcon Precision Agriculture	SGIS	ArcGIS-based data and image storage, data analysis and manipulation. VRA maps
AgLeader	SMS Advanced	Imports multiple yield file formats. Comprehensive data and image storage, data analysis and manipulation. Linked to mobile software. PA-specific VRA map formats
CaseIH	AFS Advanced	Comprehensive data and image storage, data analysis and manipulation. Linked to mobile software. PA-specific VRA map formats
Delta Data Systems	AGIS	Imports multiple yield file formats. Comprehensive data and image storage, data analysis and manipulation. Linked to mobile software. PA-specific VRA map formats
ESRI	ArcGIS	Comprehensive data and image storage, data analysis and expansive manipulation options. VRA maps as Shapefiles (.shp)
John Deere	Apex Farm Management	Farm information storage and mapping, spatial data management and analysis, image importation and wide range of PA VRA maps
MicroImages	TNTmips	Comprehensive data and image storage, data analysis and expansive manipulation options. VRA maps as Shapefiles (.shp)
Manifold	Manifold	Comprehensive data and image storage, data analysis and expansive manipulation options. VRA maps as Shapefiles (.shp)
Pitney Bowes	MapInfo Professional	Comprehensive data and image storage, data analysis and expansive manipulation options. VRA maps as .dxf or mif files
SST	SSToolbox	Imports multiple yield file formats. Comprehensive data and image storage, data analysis and manipulation. Linked to mobile software. PA-specific VRA map formats

These tasks are difficult for software companies because the data that farmers gather about variability is not always the same, and management responses differ based on local conditions and management. Some task-oriented data management software identified earlier will help with spatial decisions using a standard set of data layers, but a comprehensive solution for all farmers and locations is some way off.

Stand-alone decision support software (Table 4.7) that is not aimed at dealing with spatial data sets is available to help in making input decisions. By using these tools to make decisions about different areas on a farm or in a field, some spatial aspect to decisions can be achieved.

Spatial decision support has been improved in software for vehicle-mounted in-season crop reflectance sensors. These software programs are task-oriented and equipment-specific. They all provide some level of decision support about the 'how' and 'how much' questions for variable-rate application of fertiliser or growth regulants (Table 4.8).

Table 4.7: Examples of stand-alone decision support software that can help make simple spatial decisions on management

Company	Software	Description
APSRU	APSIM	Predicts crop yield based on climate, crop, soil and management
	WhopperCropper	Predicts crop yield at a district level using a simulation database
	HowWet?	Estimates water storage and nitrogen mineralisation
	HowMuch?	Estimates crop yield from available water and match sowing windows
Birchip Cropping Group	Yield Prophet	Makes web-based crop yield predictions based on APSIM
Back Paddock	SoilMate	Makes fertiliser recommendations based on soil and plant tests
CSBP	NU Logic	Makes fertiliser recommendations based on soil and plant tests
CSIRO	Yield + N Calculator	Calculates crop yield and nitrogen requirement, using rainfall
	Lime and Nutrient Balance Calculator	Calculates lime, nitrogen and phosphorus required to reach target pH and yield
	MaNage Wheat	Offers crop yield and protein scenarios under varying water and soil conditions
Department of Agriculture and Food WA	NP Decide	Calculates fertiliser, rate and timing for cereal crops
	OptLime	Aids soil acidity management decisions including lime quantity, and potential crop planting decisions
	Potential Yield Calculator (PYCAL)	Calculates sowing, crop nutrition and pest control based on seasonal moisture
	Yield Calculator	Offers information on yield probabilities
	Select your Nitrogen (SYN)	Aids decisions on rotation, varietal choice, tillage and nitrogen fertiliser timing, sources and rates
	WA Wheat	Advises time of sowing, nitrogen fertiliser application (timing and rate) and variety selection. Offers likely outcomes, given historical seasons
	Potassium in agricultural systems (KASM)	Advises on potassium fertiliser applications and long-term planning
eAgribusiness	MaxiYield	Advises nutrition requirements based on soil, crop and moisture
ICASA	DSSAT	Predicts crop yield based on climate, crop, soil and management
Incitec Pivot	Nutrient Advantage	Makes fertiliser recommendations based on soil and plant tests
Agri-Science Queensland, DEEDI	Rainman	Determines patterns in historic rainfall and predicts seasonal rainfall
	Soil Nitrogen Calculator (Nitrogen Books)	Advises on nitrogen management based on tests and season
Sentek	irriMax8	Soil moisture and salinity tracking for irrigation scheduling

Table 4.8: Examples of software for spatial decision support available with in-season crop reflectance sensors

Company	Software	Description
Holland Scientific	Crop Circle	Monitors crop variability and provides fertiliser distribution recommendations
Ag Leader	OptRx	Monitors crop variability and provides fertiliser rate and distribution recommendations
Trimble Navigation	GreenSeeker	Monitors crop variability and provides fertiliser rate and distribution recommendations
Yara International ASA	N-Sensor (ALS)	Monitors crop variability and provides fertiliser rate and distribution recommendations
Topcon Precision Agriculture	CROPSPECc	Monitors crop variability and provides fertiliser rate and distribution recommendations

Web-based software or services

With the improvement in internet communication speeds, greater familiarity with computer use on-farm and advances in software development, there is a growing number of companies offering web-based access to mapping and data processing tools and web-based facilities to upload farm data for contract processing. These models of data handling and analysis mean that it is not usually necessary to purchase extra software; only the proprietary task-based software delivered with a piece of PA hardware or the introductory-level data processing software provided to extract raw data from yield monitor equipment is required.

The majority of these companies are developing proprietary processing software with a varying degree of sophistication. For individuals or agronomic service providers who do not wish to deal with processing and analysis of PA data in-house, these options should prove attractive if the end products are well produced and relevant to crop management.

Key points

- Each farmer or consultant will have individual requirements for software. At a minimum, prospective software buyers should prepare a detailed list of their current requirements, and consider options for the future.
- Software for PA has been developed for farm record keeping, crop scouting, data collection, data management and analysis, creating maps, defining management classes, input decision support, navigation/autosteer, and variable-rate application.
- Software is used for both mobile and desktop platforms and different platforms may need different software.
- Software for PA is available in different levels of complexity and for different uses. The cost generally rises with the range of tasks the software can perform:

  - task-based – aimed at performing a single task or a small number of specific tasks;
  - data processing packages – data management, analysis, action and archiving in a single program:
    - introductory: data storage and simple map display options;
    - intermediate: detailed mapping and some data analysis and prescription export capabilities;
    - advanced: full mapping and data analysis and prescription export capabilities.
- Software for PA is available as freeware, shareware and commercial packages. Free/shareware options are free or cheap and are mostly aimed at data handling procedures or map viewing. Commercial packages require a purchase/licence fee but are generally user-friendly and well-supported.
- For most PA users, the intermediate- and advanced-level PA/geographical information systems (GIS) programs provide at least basic image viewing and handling tools. Some of the advanced PA/GIS software packages provide functions that allow imagery to be more easily incorporated with other data in agronomic analysis.
- Dedicated decision support packages are growing in number. Programs for specific decision tasks are available but may require a higher level of computing knowledge.
- The ability of advanced-level PA data processing packages to perform useful map algorithm procedures and other advanced data manipulation continues to expand.
- Like all software, features and compatibility issues continue to change. Software needs to be compatible with proprietary data file types used by on-farm hardware.
- Data storage and software use do not have to be done on-farm. Consultants can provide these services to growers.
- Web-based GIS are available for information transfer between grower and consultant and can be centralised with the consultant. The advanced-level PA/GIS data processing packages offer a number of alternatives.

5

Data management for Precision Agriculture

The navigation, sensing and monitoring technology being used in PA is producing an increasing amount of spatial data for farms. Like traditional financial or production data, it needs to be well-organised, labelled and stored where it can be found. All PA data processing software packages will store data in a manner that suits the operation of the particular software. This is understandable, but as the amount and types of data for a farm begin to build, it is useful to know about the different data types. Establishing and maintaining a separate storage system for the original data files will help the user understand what data is available, ensure the integrity of the data for other analysis, and provide a back-up process.

Data management

New sensors and new technologies generate lots of data. Like financial records, this agronomic data needs to be managed correctly if a farmer is to reap the benefits from the information contained in the data. If data is lost, misplaced or corrupted then any potential information value is also lost. There are several possible ways of managing agronomic data. For example:

- it could be hoarded on a computer and the mess solved once a year (not recommended);
- management of the data could be outsourced to a consultant who looks after the storage and analysis of the data and provides recommendations;
- a system of data management could be set up on-farm, allowing local access to the data plus quick and easy sharing of relevant data with others, such as agronomists.

The information presented here does not deal with analysis of the data, but with the types of data files found in PA and how to organise data so that:

- the data is not lost;
- the data is easily retrieved later.

Types of data files

There are many different types of data files used in PA. The type of file can be identified by the 'extension' which follows the 'dot' in a file name. For example, the file 'harvester.doc' is identified as a Word document by the '.doc' extension. Before trying to open a file, its type can be determined in a number of ways:

- highlighting the file with a single left click and hovering the cursor over the file;
- right-clicking on a file icon and selecting 'properties';
- using the 'view' menu in Windows® and selecting 'details'.

Different file types are developed for different purposes, so software programs will not read all types of files. They can be generally classified as generic (open) files, proprietary files or spatial files.

Generic or open files

These files can be easily accessed by common generic software, such as word-processing, spreadsheet, text editors and image editors. The file format either follows a known international standard which most common software programs accept, or is specific but with details that are freely available and allow others to write programs that can read the file format (e.g. '.txt', '.csv', '.doc', '.jpg', '.bmp', '.tif').

Proprietary files

These files are usually associated with a particular brand of machinery or sensor. They are generally used to store data recorded from specific sensors in an efficient, compact format which can later be read by designated software. The format of proprietary files is not publicly known and these files are usually incompatible with common generic or geographical information system (GIS) software. Most yield monitor software uses a proprietary file format for internal data storage and operations (e.g. John Deere® = '.gsy', AgLeader® = '.ilf', CaseIH® = '.vyg'). Not all manufacturers' software will recognise the files of other brands.

Spatial files

These files contain location coordinate information that can usually be displayed by all PA mapping, data processing or GIS software. This category includes vector files (point, line or polygon data files) and raster image files (satellite or aerial imagery). These files are often compound files, in that they may comprise several different files. There is a principal file that contains the data, and usually an auxiliary file that stores metadata including information that is used by software to project and display the data.

Vector files
These files store points, lines and area outlines (polygons). The most common is the 'shape file'. A shape file is a compound file. The header (or command) file is labelled '.shp'. Data such as the coordinates and dimensions of points, lines and vertices of polygons is contained within a '.dbf' file. A shape file has several different auxiliary files which are necessary to completely define the spatial features. These may include '.prj', '.sbn', '.sbx' and '.shx' files.

Raster image files
Remotely sensed images are stored in these types of files. Unlike a basic image file (e.g. JPG), these files have associated information about where the observed locations are on Earth. There are many different file formats; the most common in PA are listed in Table 5.1. GeoTIFF, GeoJPG and Imagine files are probably the most widely recognised by PA software. KML files are used to display data and images in Google Earth/Maps/Mobile.

The data within all these geo-referenced image files can be converted to generic text files by image processing software.

File organisation and storage in PA software

Most mapping and data processing software used in PA will have a user interface that provides access to yield and other data in a hierarchical structure similar to that shown in Figure 5.1. Users will be able to control naming at the client, farm

Table 5.1: The most common geo-referenced image file formats and associated primary file extensions

File format	Primary file extension
Band interleaved by line	.bil
Enhanced compressed wavelet	.ecw
Earth resource mapping satellite image	.ers
ERDAS Image	.img
GeoJPEG	.jpg
GeoTiff	.tif
Hierarchical data format	.hdf
Keyhole markup language	.kml or .kmz (compressed)
Portable network graphics	.png
Tagged image file format	.tif or .tiff

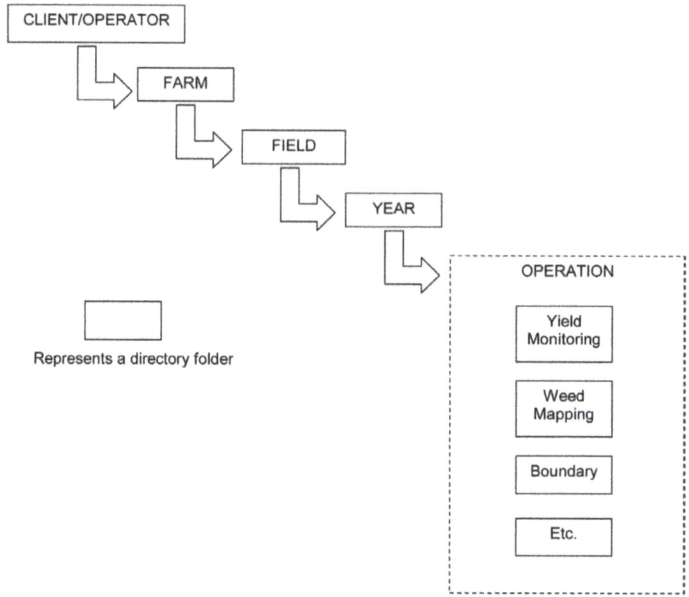

Figure 5.1: General hierarchy for PA data organisation and storage.

and field level, but often the names below this level will be controlled by the software or restricted to a number of predetermined choices available as the data is uploaded. Time should be taken to understand how file-naming and folder organisation is set up in a chosen software package, to minimise the chance of mixing up the allocation of data files.

The data in commercial software packages will not actually be stored in the hierarchy shown in Figure 5.1, but in designated back-end folders. Software designers build storage folder structures that make it easy for the software to operate and be easily updated. The names attached by a user, and visible in the menu, are used as pointers to files stored in the designated storage locations. The type of file in which the data is stored is determined by the software.

Manual file organisation and storage

Because the PA software programs do not store the data in an easily identifiable structure, it is advisable to set up another storage hierarchy outside the software storage process. It is recommended that this be done for the original data gathered for a farm.

Manual file organisation and storage involves creating a folder and file storage hierarchy similar to that shown in Figure 5.1, on a computer hard drive. As data is collected, the original files are copied into their identified location in this folder system as well as being read into the desired PA software. It is also a good idea at this stage to export any proprietary format data back out of the PA software in a generic (e.g. '.txt' or '.csv') or spatial (e.g. shape) file for storage in the manual system.

When the data is stored in this manner, there will always be one set of data that can be accessed without specific PA software, imported into new PA software and shared between users with different software. Using more farm- or enterprise-specific file and folder naming protocols in the manual system can improve the storage and retrieval of data.

File naming protocol

Correctly naming files is perhaps the most important step in avoiding the loss of data. Files that are named systematically and meaningfully will always be easy to find, or searched for if misplaced. Ideally, a name for a generic or spatial data file should include:

- field identifier;
- data type identifier (e.g. wheat, EC_a, elevation);
- year collected;
- data level (original, cleaned, analysed).

Some examples are:

- an original generic file of wheat yield data from Home field harvested in 2012 could be named Home_wheat_12_raw.txt;
- an original generic file of wheat yield data from Home field, harvested in 2012, cleaned and trimmed of outliers and with locations converted into Eastings and Northings could be named Home_wheat_12_MGA_tr.txt.

Abbreviations can obviously be used to shorten the file name, e.g. 'wht' for wheat and 'can' for canola. Naming will involve personal preferences but including references such as the above information in the file name allows the contents to be easily identified or searched for using meaningful tags.

Folder and file organisation

Folder organisation also involves some degree of personal preference, and the aim should be to ensure that the data for each field and year is kept separate. The folder for each year in Figure 5.2 has been separated into 'Original Data' and 'Modified Data'. In this instance the original data would be original proprietary files and files exported in generic format from PA software (e.g. '.txt', '.csv'). If the original data is modified then a new file should be created and saved into the 'Modified Data' folder. In this way the original data is not modified or corrupted, and it can be retrieved if analysis needs to be redone or an improved analysis applied in the future.

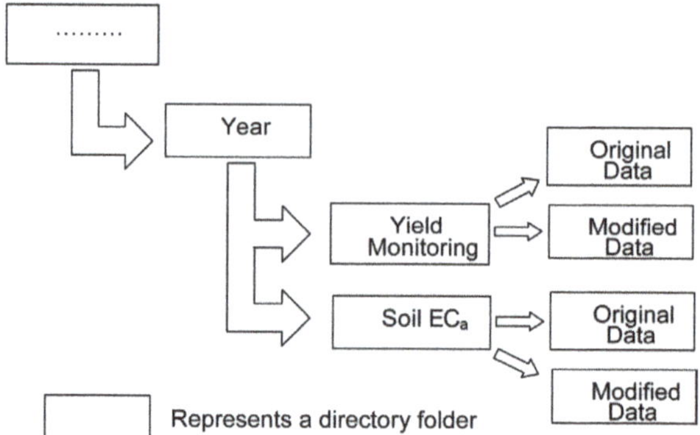

Figure 5.2: Extended hierarchy for PA data organisation and storage.

Data back-up

Like all computer files, PA data needs to be backed-up and stored in a separate location to avoid data loss from accidental deletion or computer failure.

There are two golden rules for data back-up:

- do it regularly and as part of a routine, e.g. every Friday at the end of the working week;
- keep the back-up copy in a separate, safe location away from the main computer.

The simplest option for data back-up is to copy data onto a portable, external hard disk. Data sticks, compact discs and DVDs can also be used. It is possible to set a computer for automatic back-up at regular periods using the operating system, some third-party software or an internet-based data back-up website.

ALWAYS BACK-UP DATA REGULARLY

If the computer hard disk crashes and the data is lost, it may be gone forever.

It is also important to regularly download and back-up information that is stored on PA consoles in farm vehicles, particularly during periods of intense use. Do not wait until the end of harvest to download all the harvest data! Regular downloading avoids data storage units becoming full and either overwriting data or not recording it.

Data ownership

Original data

Original data ordered or gathered by a grower should remain the property of the grower. Following contracted surveys, such as an EC_a soil survey or remote sensing of a crop, the raw data as well as any maps or other processed/derived products should be provided by the contractor. Having hard copy maps is useful to a grower but they limit how the data can be used in the future, even assuming that the map has been well constructed. When yield data is collected by a harvest contractor the situation is a little less clear. Depending on the contractual arrangement, the contractor may retain the right to sell the data to the grower.

Processed data layers

Ownership of processed data can be more problematic. When the expertise of a contractor has been used to transform the raw data (e.g. a yield map or a biomass image) into information (e.g. a prescription fertiliser map), data ownership still rests with the grower but the contractor may negotiate to retain some rights. Negotiation on any external use of the processed data should also be considered.

Companies have begun providing internet-based services to growers for data mapping and interpretation. Growers can upload data and access tools to make a range of map products (e.g. variable-rate application maps). However, to do this the growers must release the raw data to the on-line service provider and therefore the ownership and use of the data need to be verified.

Data pooling

While there may be the possibility of exploitation by parties involved in a system of pooling data, there are potential advantages to growers and the grains industry from aggregating individual data archives. These include:

- creating local bench-marking tools from the pooling of spatial production data (e.g. crop yields) to help identify production issues;
- sharing on-farm experimental data for a local area. This can greatly multiply the number of treatments and/or replicates for a more thorough analysis of alternative management approaches to locally identified issues;
- fine-tuning of yield and other production goals;
- gaining a clearer picture of production variability on farms in the region for supportive agricultural businesses;
- easier identification of commercial opportunities.

There are three general organisational models that could be used for data pooling:

- those run by agricultural manufacturers or input suppliers, which may benefit growers using particular types of machinery or products;
- those run by independent data management companies, which may be agronomists or linked to an agronomist;
- those run by non-profit data management groups, which may include local grower groups or government agencies that provide a specific local/regional focus.

Key points

- If data is not collected or it is lost, it cannot be converted into information and decisions.
- Data should be downloaded and backed-up regularly as part of good PA management.
- Data will come from different sources and in different file types. Generic, proprietary and spatial files may all be used in a PA operation.
- An understanding of which file formats can be handled by particular PA software is important, to minimise problems with data transfer.
- Data files should be named in a systematic fashion to allow easy identification and retrieval.
- Original data files should be stored unaltered. Data should be saved to a new file before data processing or analysis is done.
- Always aim to obtain or provide the raw data as well as any derived data layers from a contracted survey.
- Most mapping software will automatically store any data that is gathered or imported. However, at the very least, all original data should also be organised and stored in a digital file system outside any specific software.
- A simple hierarchical file storage system should be used that relates to the farming operation (e.g. farm/field/year/operation). In this way data is easy to find and share. It is also easy to back-up to external storage.
- Growers should retain the right to any original data they collect or for which they pay.
- Data pooling may help growers in a region benchmark, identify optimal differential management strategies more quickly and fine-tune production goals.
- Data pooling may provide an insight into local variability and potential commercial opportunities for local agribusiness and consultants.

6

Making and interpreting maps for Precision Agriculture

Yield monitors are now standard on many new combine harvesters. Instruments that gather data about soil properties and other crop attributes are also widely used. Coupling these monitoring systems with global navigation satellite system (GNSS) technology allows growers to gather location information with their data. However, to make decisions from this information the data needs to be presented in a form that is easy to interpret. Making continuous surface maps helps interpretation by visually displaying patterns of variability in the data. However, if yield, soil or other production maps are incorrectly constructed and/or displayed then making decisions from them may be difficult, or incorrect. There are many options for making and displaying data.

Making maps from PA data

The process of making geographical maps is known as cartography. In PA there are two general types of geographical maps:

- reference maps – these use general information about boundaries, natural phenomena and infrastructure to provide a relative point of reference to activities. An example would be a map using an aerial photograph with farm boundaries overlaid;
- thematic maps – these display information about a single theme in relation to a specific area. An example would be a crop yield map for a field.

This chapter will concentrate on creating thematic maps from PA data files to represent how an attribute of interest may be changing over a field or farm. Thematic maps used in PA can be based on two distinct types of data:

- point data – data sets that match a measured value of an attribute at a specific point in a field with information about where the measurement was made. A sensor provides the measured value and a GNSS provides the location coordinates. An example would be data supplied by a crop yield monitor;
- raster data – data sets that provide a measured value of an attribute from each cell or pixel in a grid of pixels covering the area of interest. The location information is based on the ordering of the pixels. An example would be a satellite image of a crop in a field.

Point-based maps

Data quality

Making point-based maps is a relatively simple process, but it is important to remember that such a field map is a representation of individual data points gathered across the field. Before beginning the mapping process, it is important to establish that the data is accurate and sufficiently detailed.

If the data has been gathered and/or stored inaccurately, then the map will be inaccurate. An inaccurate map may lead to poor management decisions. Yield maps are the culmination of a season's work and a field cannot be reharvested. Thus, attention must be paid to the whole data gathering and storage process. Other monitoring operations, such as EC_a surveys, can be done again if necessary, but at a further cost of money and time.

Remember: with all crop and soil monitoring it is important to ensure that enough time is taken to gather the best quality data possible.

Cleaning raw data files

Ideally, all raw data files should be checked prior to making a map. Values that are obviously too high or low should be removed. The thresholds will depend on what is being measured, but in almost all cases, zero or negative values and those considered excessively high for the location should be investigated for removal. If there are errors in GNSS operation, some values may be matched with location information that is obviously wrong; these should be removed. For many types of data, these steps can be done in a spreadsheet program by plotting the data using the location information, then exploring the maximum and minimums of the data.

Crop yield monitoring systems often automate this process in the software used to read the files from the harvester storage media. Each system allows a different degree of control over what information can be used to identify bad data, and the level of any cut-off parameters. It is important that these features be used where present. All the advanced-level PA data processing software allow users such control.

Presenting the data

Maps are the easiest way to grasp the amount of variability in a measured property, and how it changes over fields. Data gathered at individual points across a landscape looks like Figure 6.1 when in its raw form. In the crop yield example, each point has information about its location and a yield value.

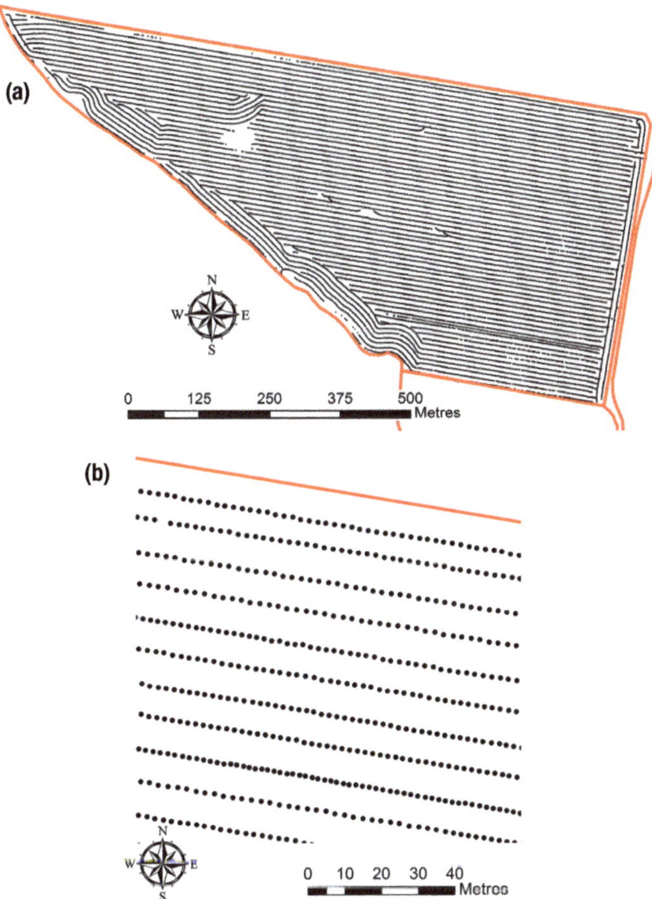

Figure 6.1: (a) An example of raw crop yield point data following basic cleaning. (b) A close-up of a section of the field showing the point nature of the data.

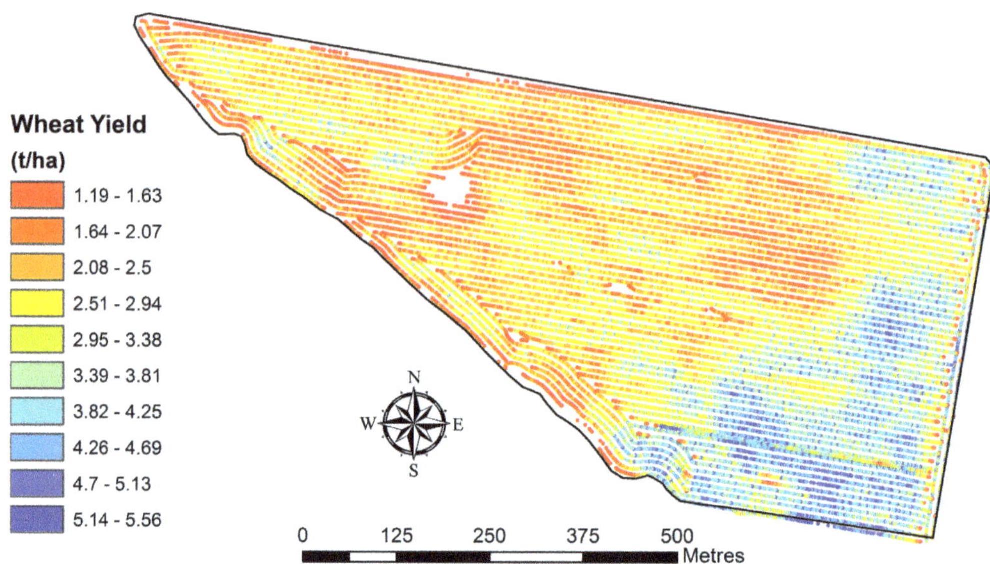

Figure 6.2: A dot map created by colouring raw point data in 10 categories.

Dot map

In many mapping software programs, it is possible to allocate the measured values of a property at each location into categories and to give each category a different colour. This produces a dot map, such as the example in Figure 6.2.

Grid map

The point data can also be made into a simple grid or raster map (Figure 6.3). The area around each data point is shaped into a square that butts up against its neighbours. The squares are filled with the colour representing the value of the measured property. In this way, all the white space between the points is removed.

These two maps provide essentially the same information. The raw information has been allocated into 10 categories based on the yield value at each point. Any irregularities from the data sampling or measurement process that remain in the raw data after cleaning will be present in these maps. To gain the most accurate picture of a field's variability, the processes involved in collecting the data must be understood. When sampling and measuring a property with a simple and accurate analysis, such as taking crop or soil samples by stopping at a point every few metres, the maps shown may well be a good representation.

On the other hand, if the point-to-point accuracy of the measurements is questionable then a smoothed continuous surface map may be more appropriate. Crop yield data is a good example of such a case as it is gathered by a large, moving vehicle using threshing, augering and elevator processes to deliver grain samples to a sensor near the grain bin for measurement each second. A harvester travelling at

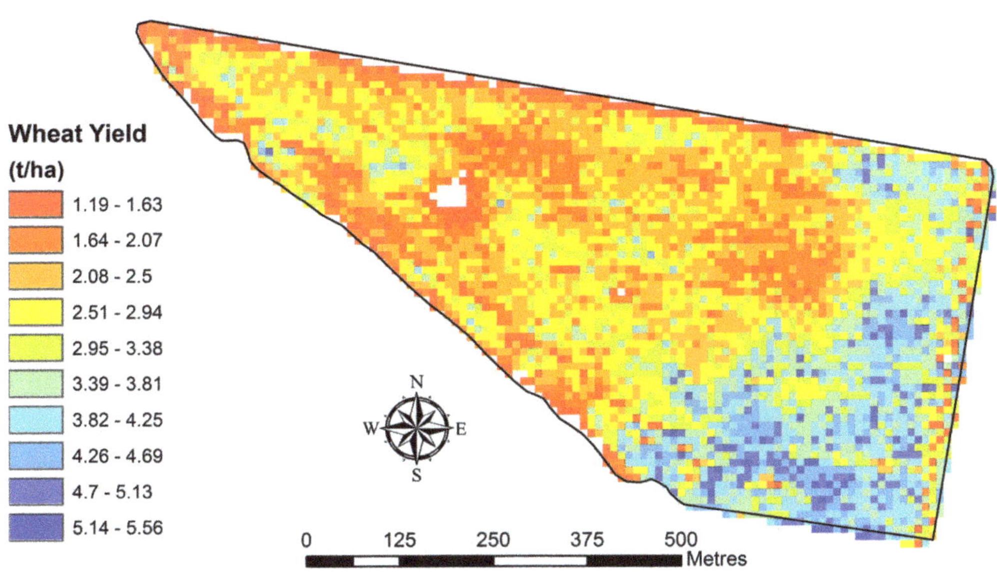

Figure 6.3: A grid map created from raw point data in 10 categories.

6.5 km hr^{-1} will cover 1.8 m per second. With a 10.7 m front (35 ft) in a 3.5 t/ha wheat crop, the yield sensor is measuring on average 6.5 kg of grain each second. While the sensors are accurate at this level, just where the grain came from along the harvest path is not so easy to determine over such short distances. Research has shown that the harvesting process can result in grain being mixed together over harvester travel distances of 20 m.

Continuous surface map

In a continuous surface map, neighbouring measurements of a property are blended together to create a smoother transition between each point. This results in the removal of the sharp distinction between measurement points seen in the dot and raster maps.

As can be seen from Figures 6.1 and 6.2, the data from most PA sensors is gathered at irregular distances along operational paths that may also be unevenly spaced. Variations in vehicle speed and driving accuracy are the main contributors to these irregularities. To overcome this, continuous surface maps are created using spatial prediction processes.

Continuous surface maps made using spatial prediction are valuable from an analytical perspective as they permit data from different times and/or sources to be compared. When a field is harvested, it is highly unlikely that yield from the same locations will be recorded in different years. This makes it difficult to merge data from different years and to perform statistical analysis. Continuous surface maps provide the opportunity to compare values at the same points in the field over any number of yield maps. This allows the statistical, rather than visual, identification of stable and variable areas of crop production. Similarly, data from other sensors, for example aerial imagery and on-the-go soil sensors, when predicted to continuous surface maps, can be included to examine yield-determining factors.

Spatial prediction process

Spatial prediction is based on the concept of spatial dependence. This is the idea that samples or measurements (observations) of an attribute taken close to each other in a field are more likely to be similar to each other than those separated by larger distances.

Spatial prediction requires a description of the spatial dependence in the form of a mathematical model. The model describes how similar the observations in a data set are, based on their relative locations. The spatial prediction process then applies this model to estimate values of the attribute at locations in the field where no observations were made (prediction points).

The process generally works by laying a fine grid of prediction points over the raw data then 'predicting' a value at each grid point, using the observations from nearby sampled locations. The nearby observations are typically chosen by defining a 'local neighbourhood' boundary around the prediction point (Figure 6.4). This is done by specifying a maximum distance from the prediction point (e.g. 50 m) within which observations will be used, or the closest maximum number of observations (e.g. 100). Choosing the neighbourhood boundary is often user-definable in PA mapping software.

The values at the new points are then allocated to yield categories and areas of similar value filled with the same colour (Figure 6.5) This whole process is

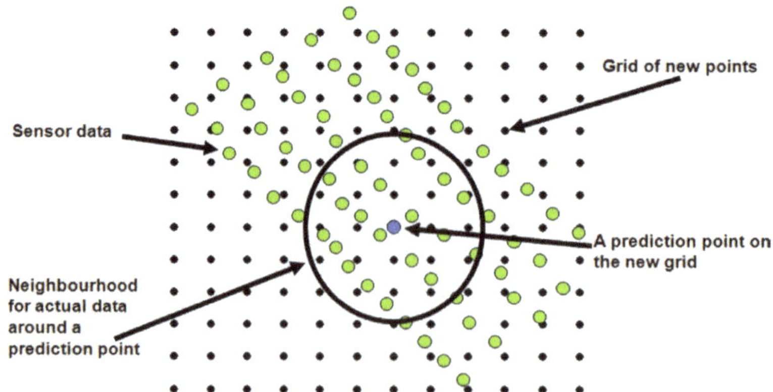

Figure 6.4: The general process of spatial prediction used to make a continuous surface map.

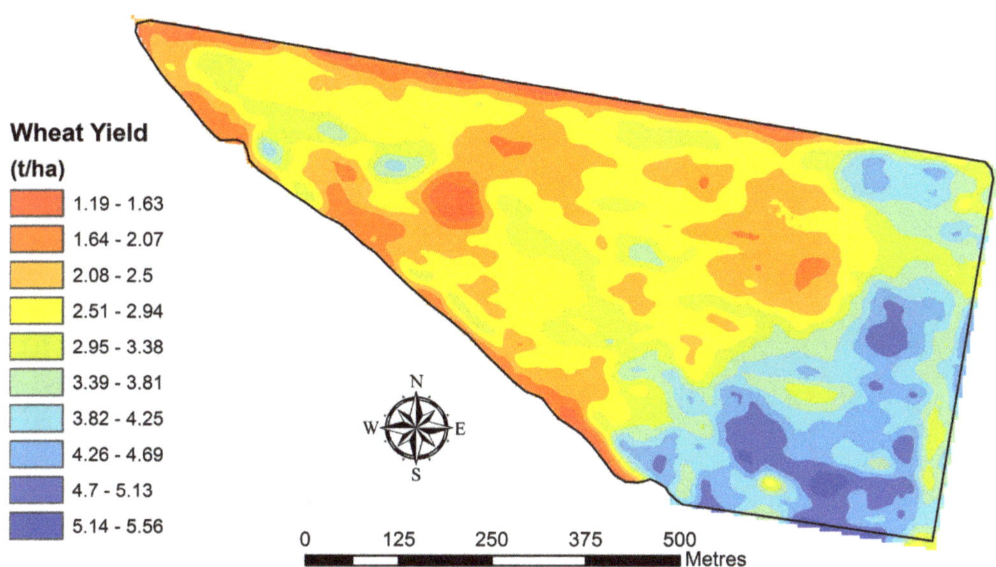

Figure 6.5: A continuous surface wheat yield map.

performed in the background in most software mapping packages when users ask for a 'contour' or 'spatial' map.

Figure 6.6 is an illustration of the stages in making a surface map, using raw crop yield data.

Spatial prediction methods

The degree to which the value from each nearby observation contributes to the prediction estimate is known as the 'weighting'. The 'weights' are calculated from the model of spatial dependence. The spatial prediction methods used in PA can be distinguished by the model each uses for describing the spatial dependence in data sets.

For the prediction process, each observation in the defined local neighbourhood around a prediction point is multiplied by its weight. The weight is expressed as a fraction (i.e. 0.6); the size of the fraction depends on the number of observation points to be used in the prediction and how the model expects each observation value and the value at the prediction point to be related. To ensure that the predictions are kept in the correct data range, the weights allocated to the observation values for each prediction must add together to equal 1.0.

There are three main spatial prediction methods for PA:

- moving mean;
- inverse distance;
- kriging.

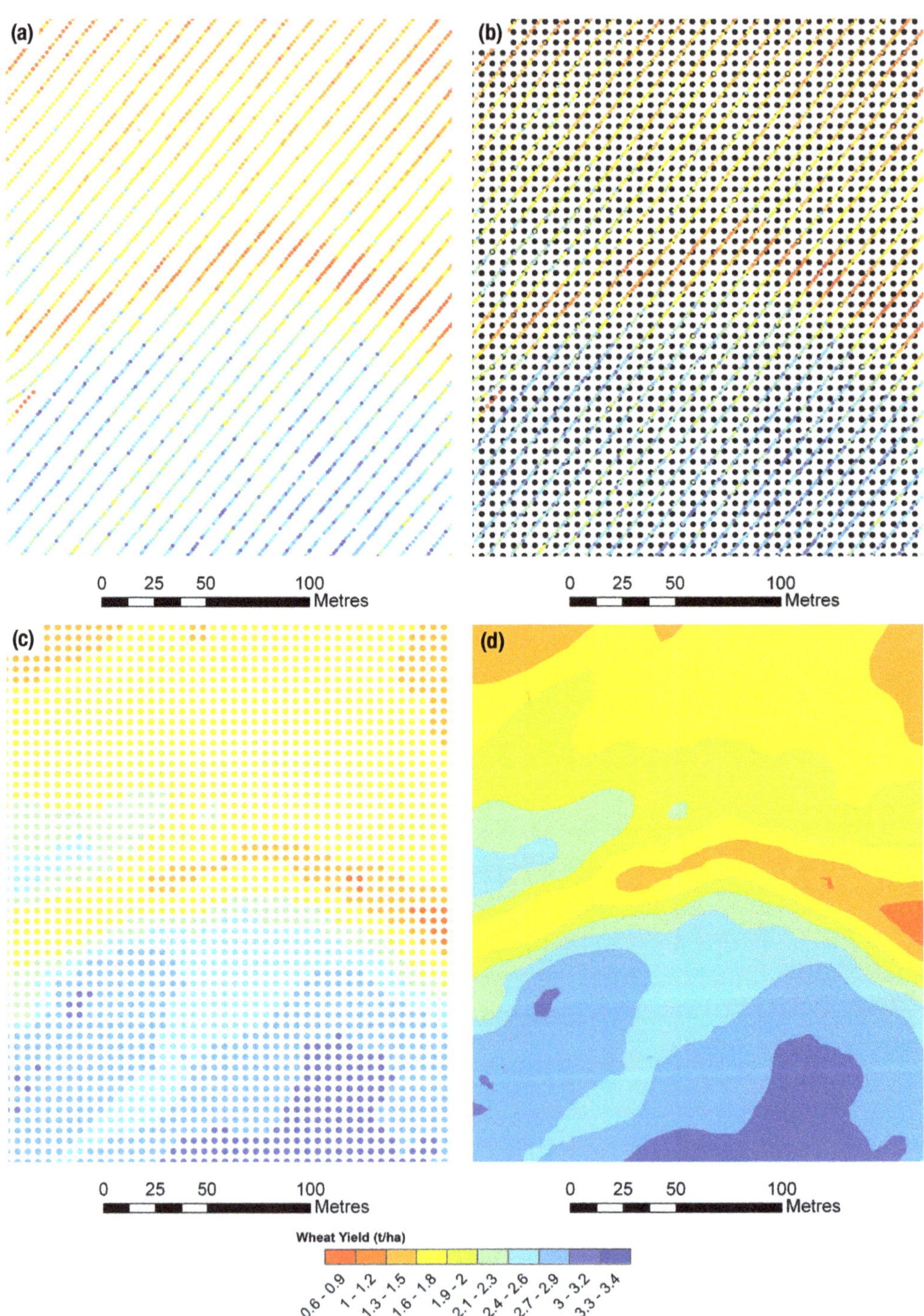

Figure 6.6: The spatial prediction process. (a) Raw crop yield data. (b) Raw crop yield data and a grid of prediction points. (c) Predicted crop yield values on the grid of prediction points. (d) An 'area filled' surface map.

Moving mean

The mean (average) of observations within a local neighbourhood around the prediction point is allocated to the prediction point.

The applied weight is uniform for all observations in the neighbourhood, which assumes all observations have equal relevance to the yield at the prediction location.

Inverse distance

This prediction process weights observations within the defined neighbourhood in proportion to their distance from the prediction point. Closer observations receive a higher weighting than those further away.

The weights are calculated on the assumption that any observations (e.g. yield, soil or reflectance) are related to each other according to a 'universal' function based on the distance between observations. No account is taken of the relationship in the actual data set being used. Commonly, the weights are calculated proportional to the square of the distance (d^2) – the process is termed 'inverse distance squared'. The effect of raising the power of the distance effect (d, d^2, d^3 etc.) is to reduce the weighting to observations in the neighbourhood that are farthest from the prediction point (Figure 6.7). With inverse distance squared, observations further than 5 m away from the prediction point are given little influence.

Some PA mapping software provides a combination of inverse distance and moving mean spatial prediction. This is to overcome situations where the inverse distance process causes the resulting map to appear 'spotty'. The user has the option of choosing a distance or ratio that applies a mean weighting for points up

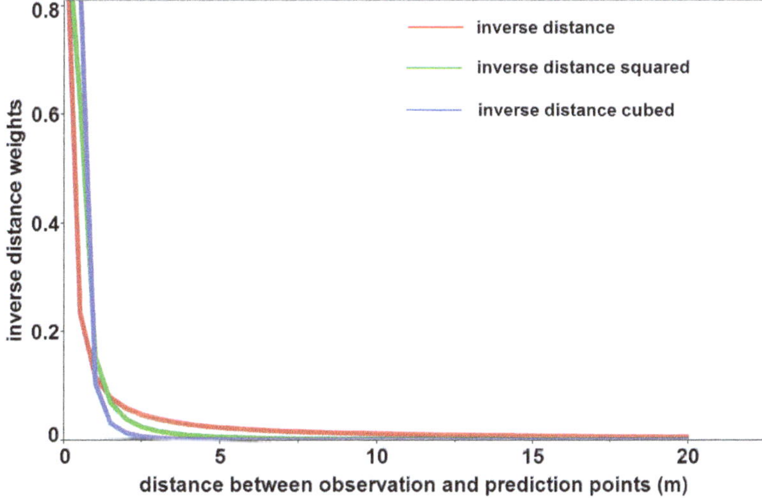

Figure 6.7: An example of the weights for prediction as determined by different inverse distance dependence models.

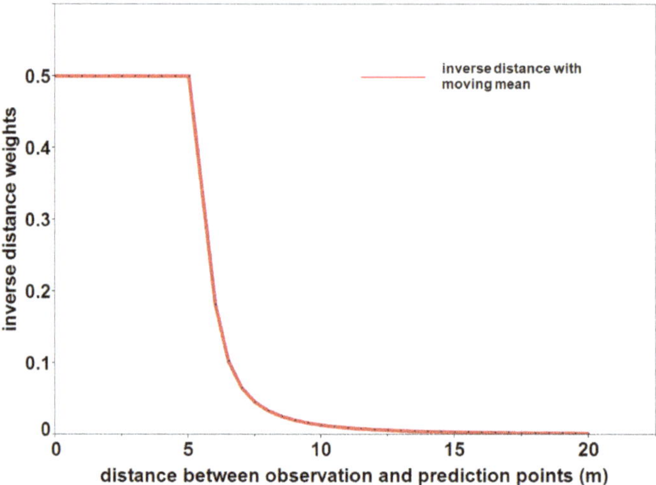

Figure 6.8: An example of weights for prediction as determined by the hybrid of moving mean and inverse distance spatial models.

to the chosen distance away from the prediction point, then applies the inverse distance effect to those observations further away (Figure 6.8). This model has the effect of smoothing the resultant map.

Kriging

Unlike the universal model used in inverse distance, kriging examines each data set to determine the best model to describe the relationship between the observations. The relationship is still based on the distance between observations, but it is tailored for each set of observations. The kriging process uses the variogram to model this relationship.

The variogram model may take a number of different shapes, depending on the data set. A generalised example of a variogram model is shown in Figure 6.9. It essentially describes the amount of variation to be found in the data as the distance between observations increases. Weights are then obtained for the neighbourhood observations surrounding each prediction point using this model in the kriging prediction process.

The variogram model can be calculated from a whole-field data set to provide a single model for use at all prediction points (global variogram). Alternatively, a variogram can be calculated in each neighbourhood for use only at the local prediction point (local variogram).

Comparison of maps

The moving mean process produces the smoothest map of the three spatial prediction methods (Figure 6.10a) because all observations are given equal weight

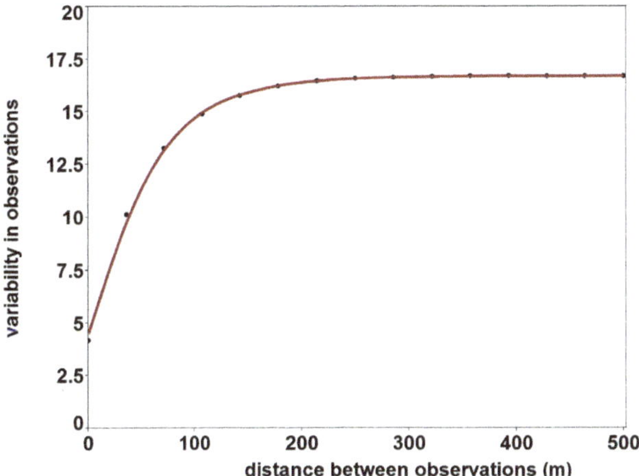

Figure 6.9: A variogram model used in the kriging process.

in the prediction. The smoothness is accentuated because prediction points in close proximity often share a number of observation points and, without some form of distance weighting, the predictions will inevitably be similar.

The inverse distance method incorporates more variation because of the distance weighting. Figure 6.10b shows a map made using inverse distance squared. The spottiness of the map is a result of the predictions being strongly influenced by observations very close to the prediction point locations.

The map made using the kriging procedure (Figure 6.10c) includes more variation than the moving mean map, but none of the spottiness of the inverse distance squared map. The influence of observations on predictions is determined by the data set, so the overall field variability is represented.

Using a combination of inverse distance and moving mean will produce a map with intermediate variation, depending on the smoothing parameters chosen. This process, or moving mean by itself, would be preferable to the inverse distance method if kriging was not an option in the available mapping software.

Raster-based maps

Raster maps are usually produced from imagery gathered from satellites or aeroplanes. Unlike the data gathered by point-based sensors, the imagery data is gathered on a cell basis to create a complete coverage of the area of interest. These cells are the smallest area of measurement in an image and are known as pixels (from the combination of 'picture' and 'element'). Each pixel represents a single data value and no spatial prediction is usually required. This arrangement is shown in Figure 6.11.

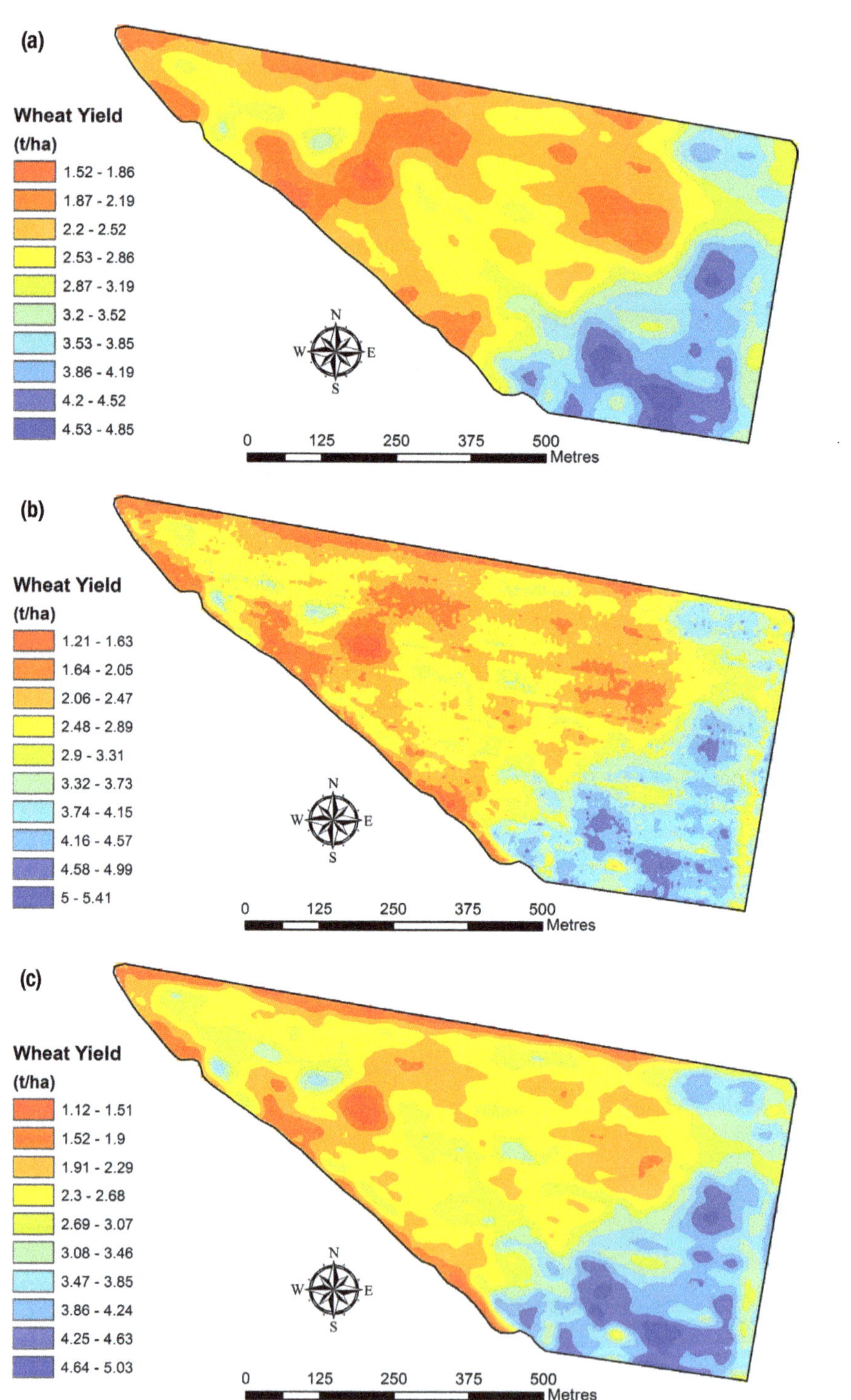

Figure 6.10: Maps of the crop yield data in Figure 6.2 made using different spatial prediction processes. (a) Moving mean. (b) Inverse distance squared. (c) Kriging.

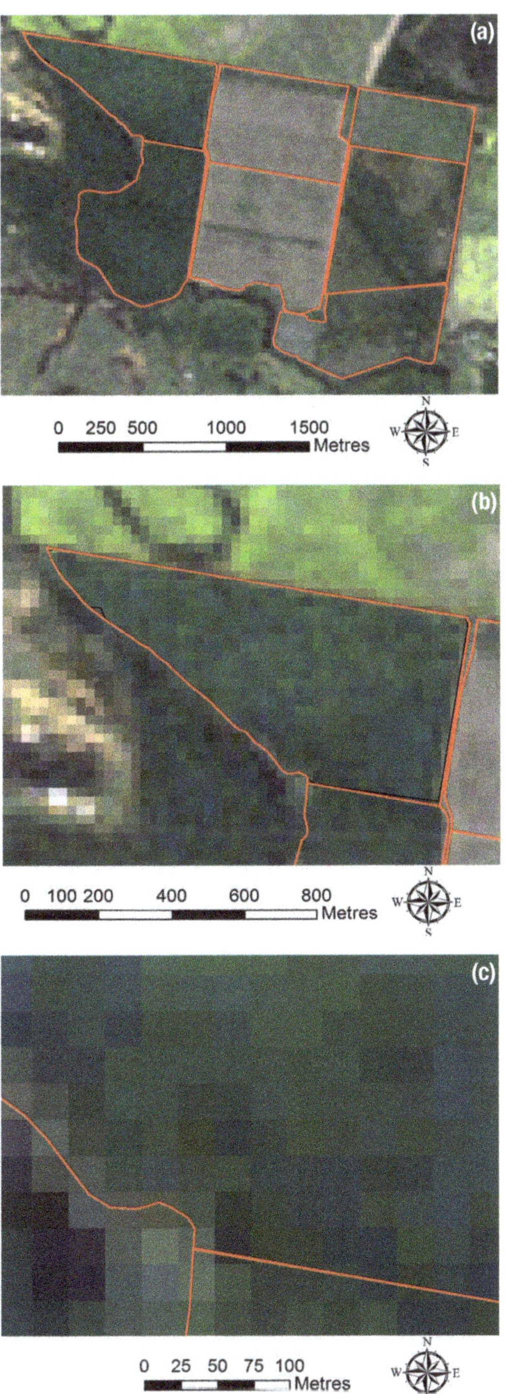

Figure 6.11: Raw raster data provided by a Landsat satellite image. (a) Farm scale. (b) Field scale. (c) Close-up within the field showing the 25 m pixels, the smallest area of measurement in the map-oriented Landsat data.

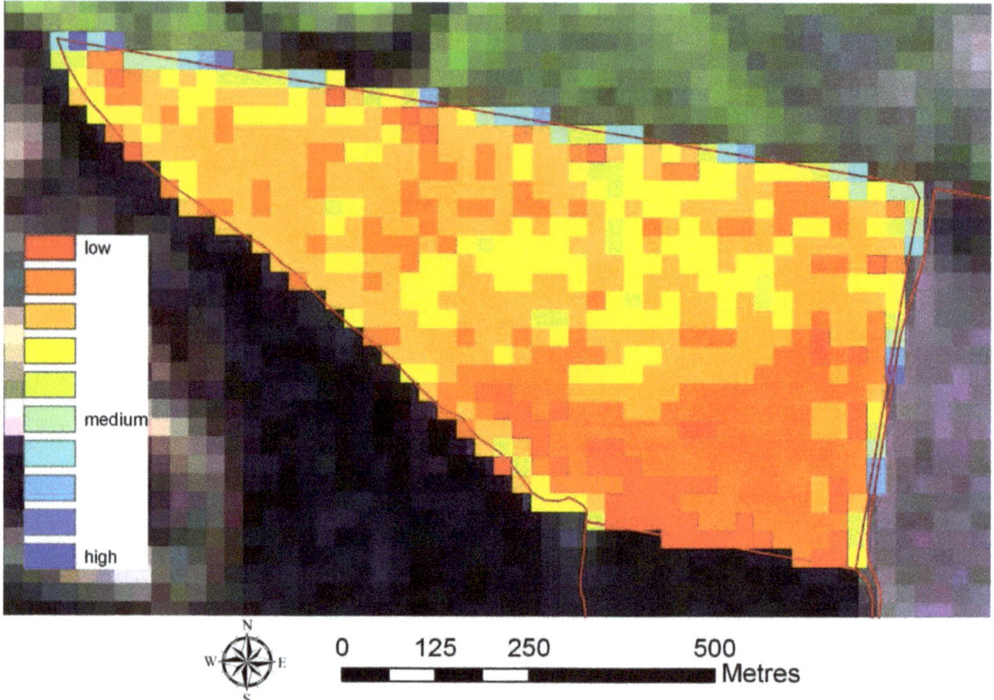

Figure 6.12: A raster map produced by applying a colour scale to the Landsat data in Figure 6.11.

The pixels can be assigned to a colour scale just like the point-based data, to produce a similar thematic map (Figure 6.12). These maps look similar to the grid maps produced using the point data in Figure 6.3.

The map legend

Irrespective of the type of data a map is based upon, all good maps must have a legend. A map legend should obviously contain the units of measurement for the property being mapped, but should also give meaning to the map symbols, colours, distance/scale and orientation.

Choosing the colour scale

The choice of colour scale is ultimately a personal decision. However, making certain colour scales standard for maps of common crop and soil properties is useful. Such standardisation helps people quickly understand what is 'high' and 'low' in a map. This improves interpretation and reduces confusion in discussions about the maps.

The most common map is a crop yield map and a standard colour scale has been proposed for use in Australia. This is the 'rainbow' colour scheme (Figure 6.13). The vast majority of software mapping programs include this scale, where red is 'low' and blue/violet is 'high'.

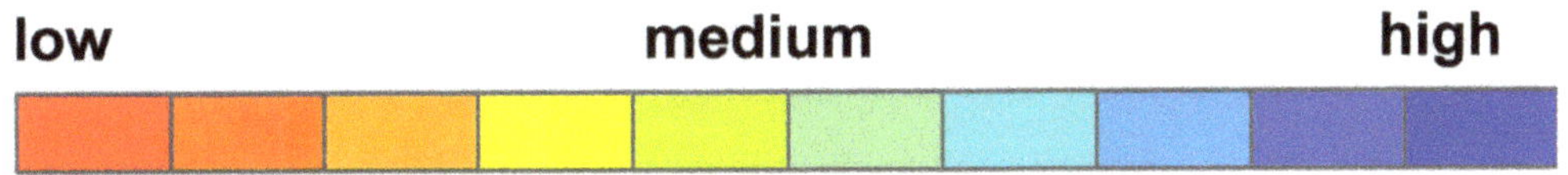

Figure 6.13: The rainbow colour scheme that is commonly applied to crop-related maps.

It is common for the rainbow scale to be used for other crop-related properties such as biomass, reflectance indices and protein content. Using this scale for maps of crop properties is recommended.

While this scale could be used for all PA maps, it is usual and preferred for non-crop related maps to use a different colour scale. This enables quick distinction between cropping and other maps (Figure 6.14).

Above all, consistency in colour scales across map sets should be maintained over farms and seasons.

Choosing the number of categories

The number of categories into which the data range is broken affects how variable or busy a map looks – too small a number and the variation in a map is flattened, too large and the variation is exaggerated (Figure 6.15). As a general rule for mapping variation in crop and soil properties, five to 10 categories would cover most uses in PA.

The main idea is to make sure that the changes in the scale reflect an important change in the property being mapped. For example, changing colours as crop yield increases by 0.1 t/ha would usually be considered as exaggerating the variability from a fertiliser management perspective.

Allocating data into categories

Most yield-mapping programs allow the user to select from a range of allocation methods. Some or all of the following may be available.

Equal interval

The data is split into a user-defined number of categories which each span an equal range of the data (e.g. 0.5 t/ha increments), so the number of points in each category will usually vary.

Quantiles

The data range is split into a user-defined number of categories that each represents a fixed percentage of the data. For example, five classes means that each category will contain 20% of the data (100%/5=20%). The categories will contain the same number of points, however, the width of the ranges will usually vary.

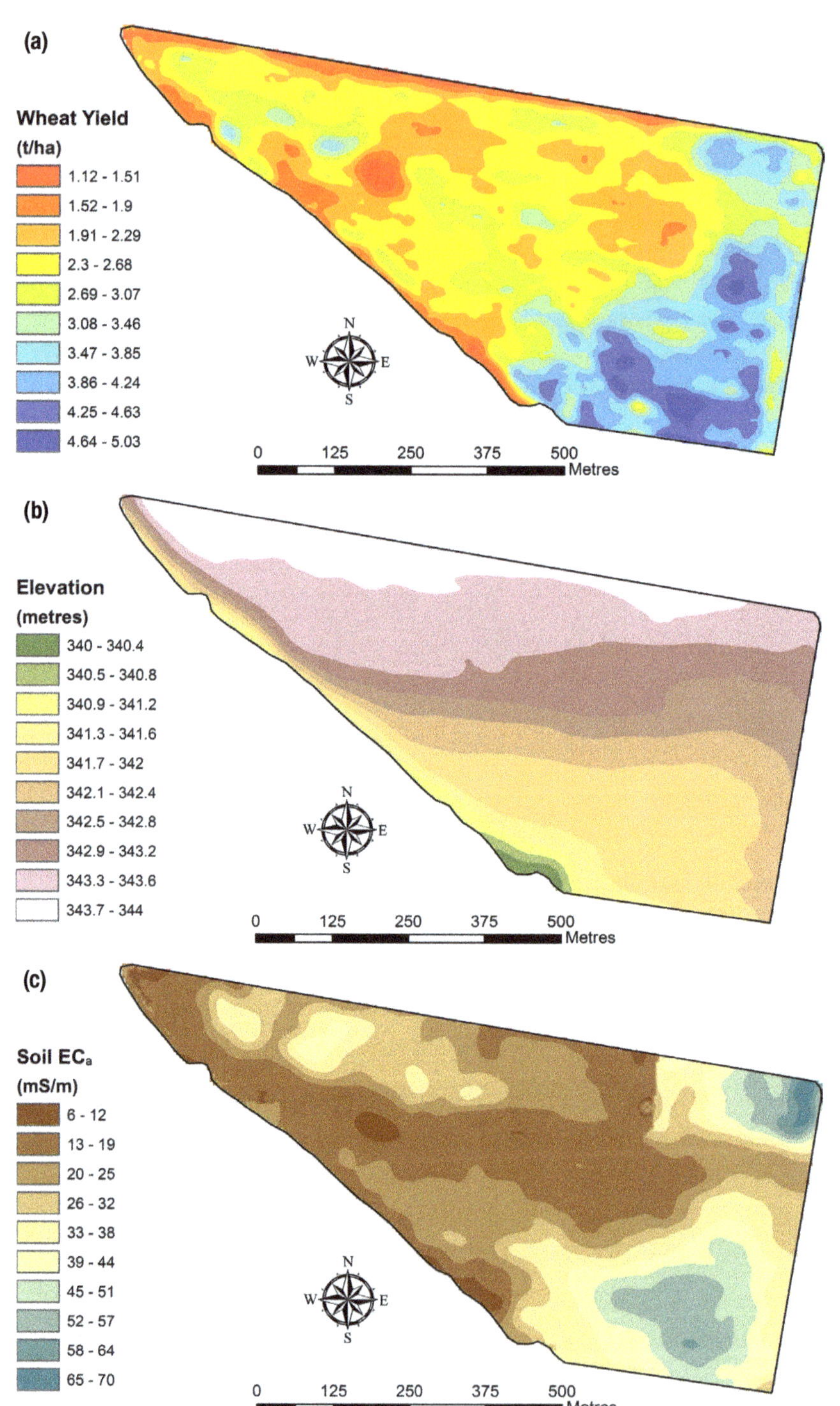

Figure 6.14: A series of maps for a single field showing the use of different colour schemes for (a) yield, (b) elevation and (c) EC$_a$.

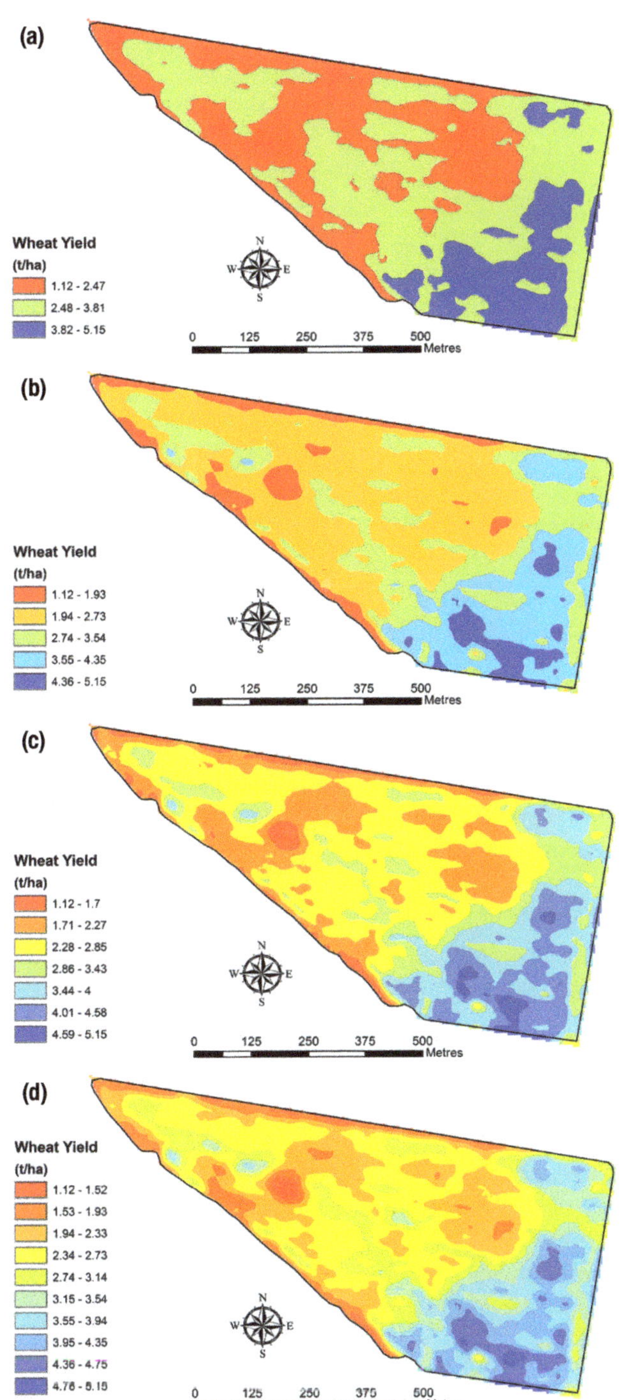

Figure 6.15: The number of categories in the colour scale affects the variability seen in a map. (a) Three categories. (b) Five categories. (c) Seven categories. (d) 10 categories.

Standard deviation

This creates categories based on the variation in the data. Categories are created above and below the overall mean, with ranges equal to a user-defined portion of the standard deviation of the entire data set. One standard deviation above the mean covers 32% of the data and an additional standard deviation category covers a further 16% of the data. The same applies below the mean, so that four standard deviations (two above and two below the mean) will cover 96% of the data in normally distributed data sets. The categories therefore span an equal range in the data, but the number of points in each will vary. More points will be allocated to the first categories on either side of the mean.

Natural breaks

This creates a user-defined number of categories based on minimising the variation within each category. The categories will contain an unequal number of points and the width of the ranges will vary.

Linear stretch

This is usually only applied to raster imagery data where the full data range is stretched over a colour gradient. It can cause difficulties in comparison if the upper and lower ends of the range are not consistent between years.

User-defined intervals

Data values are allocated into categories that cover ranges as defined by the user. The category ranges need not be equal and the number of points will usually vary.

Interpretation of allocation methods

There are a number of implications for visual interpretation when using each of these allocation methods. See Figure 6.16 for a visual comparison of the effects.

Equal interval

As shown in Figure 6.16a, this method can hide variation if the yield ranges are not scaled to fit from the minimum to the maximum of the data being mapped. As long as the scaling is correct, it is easy to set the category range to a figure of agronomic significance so that variation is easy to interpret and compare across maps.

Quantile

Figure 6.16b shows that this method can exaggerate variation because it puts the same number of points into each category, so the low and high categories have larger ranges than with other methods.

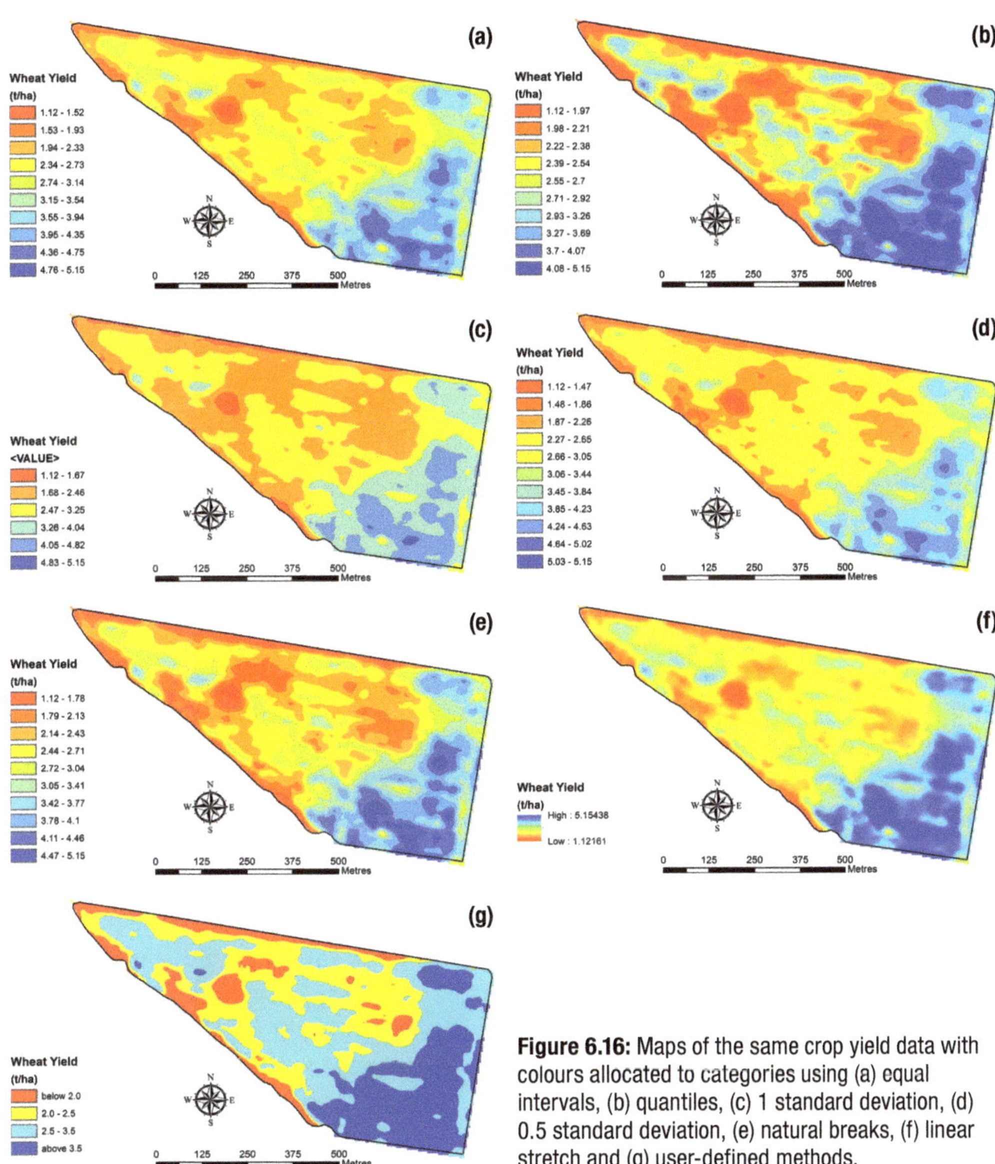

Figure 6.16: Maps of the same crop yield data with colours allocated to categories using (a) equal intervals, (b) quantiles, (c) 1 standard deviation, (d) 0.5 standard deviation, (e) natural breaks, (f) linear stretch and (g) user-defined methods.

Standard deviation

Figures 6.16c and 6.16d show that standard deviation tends to smooth out variation as it allocates more points to the categories closer to the mean for the field. This results in the higher categories having smaller ranges, so they appear less prominent.

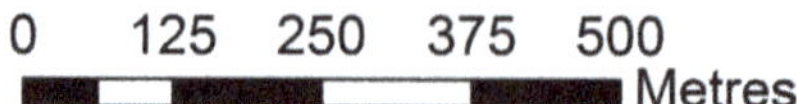

Figure 6.17: A distance scale bar used to establish the physical scale of the mapped area.

Natural breaks

As shown in Figure 6.16e, this appeals when the aim is to make categories where the included points are as similar as possible. This process appears closest to the equal interval procedure when the number of categories is reasonably high (>5). However, this allocation will not be consistent from map to map, making comparisons less simple.

Linear stretch

Figure 6.16f shows how this exaggerates the variation. Like the quantile method, it gives equal importance in separation to the low, mid and high values.

User-defined

This allows custom maps to be made when looking for particular cut-offs in variation (Figure 6.16g). User-defined maps are unique.

To compare maps over time, it is necessary to use a consistent allocation methodology. The different methods described here tend to enhance different sections of the data. For the clearest interpretation and comparison of crop yield maps, it is advisable to use the equal interval allocation method.

The distance scale and orientation indicator

All maps should include an indicator that allows a viewer to establish the physical scale of the area being mapped. This is usually done by including a distance scale bar (Figure 6.17), but it may also be achieved by including a Cartesian coordinate frame (Figure 6.18).

All software mapping programs should display maps with 'north' represented by the top of the screen or page, but to remove potential confusion a 'north arrow' or indicator of the direction of the cardinal points should be provided.

Interpreting PA data maps

There are three basic rules for interpreting PA data maps:

- natural variation usually appears as an irregular pattern. Any large changes will have a reasonable gradation between the different levels;
- linear patterns are not usually the result of natural processes. In nearly all cases they reflect management operations (e.g. old fence lines, sowing/fertiliser/agrochemical application problems and machinery errors) or the effect of underlying infrastructure (pipes etc.);

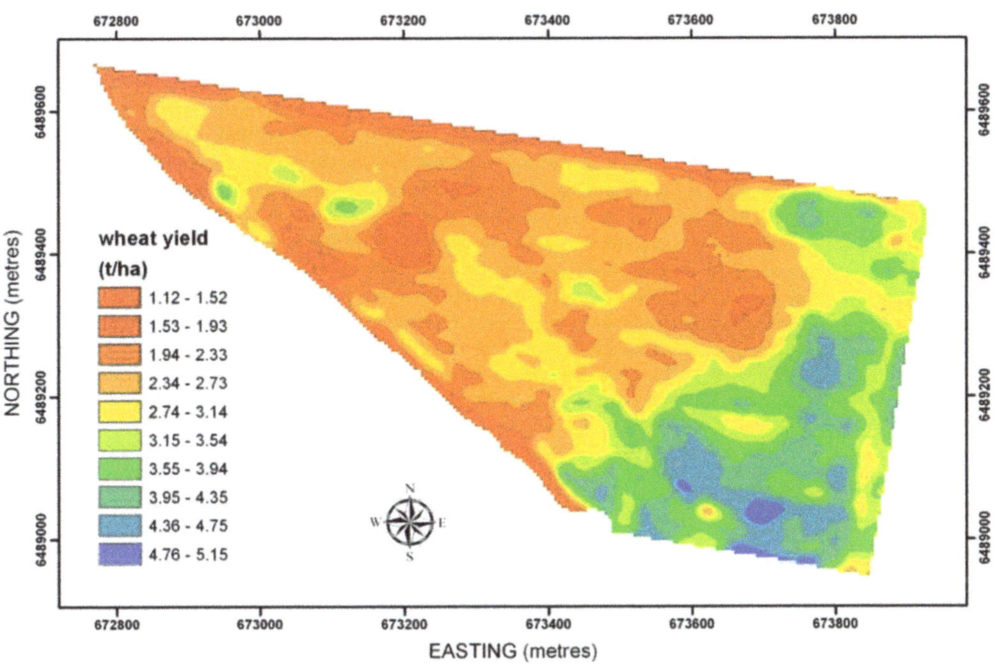

Figure 6.18: To provide information on the physical scale of the area being mapped, a Cartesian coordinate frame may be included instead of a distance scale. A Cartesian coordinate system uses distance, not degrees.

- irregular patterns that display abrupt changes are often the result of management operations (e.g. chemical spray drift, partial field irrigation) or sharp topography changes.

Identifying any maps, or sections of maps, that are affected by technical, operational or catastrophic external influences is crucial prior to using map data in any analysis for SSCM.

Interpreting yield maps

Discriminating between natural variability and anomalies

Yield maps are often the first type of PA map encountered and the degree of variation and patterns they express can be initially overwhelming or confusing. Yield maps are also widely used to assess outcomes from SSCM operations. These are good reasons for being able to distinguish between natural and induced variability, to identify potential causes of anomalous variation and to hone in on important issues for future management intervention. Table 6.1 provides a guide to identifying and interpreting yield map anomalies.

Figure 6.19 shows a yield map with a number of features which have been caused by management operations. Changes to field boundaries often mean that areas with different management histories are combined into a neighbouring field. The effects are usually obvious as linear features (a) in the first yield maps, but may

Table 6.1: A guide to identifying and interpreting yield map anomalies

Straight-line effects		Irregular effects	
Along working direction	**Across working direction**	**Curved or wandering lines**	**Patches**
Equipment or application errors during: • sowing • fertiliser application • agrochemical application	Previous working lines Roads Soil compaction	Poor fertiliser application Poor soil ameliorant spreading	Changes in soil type Changes in soil chemistry Changes in soil physical properties
Different varieties	Old fence lines	Contour banks	Weed infestation
Different sowing dates Different sowing conditions	Underground: • pipes • drains • utilities	Natural changes in topography	Natural topographic changes leading to: • waterlogging • frost damage
Application of different chemicals		Agrochemical spray drift	Impacts of previous management Impact of previous crop productivity
Application of different fertiliser/rates		Pest damage on borders	Disease impact Animal damage Insect damage
Soil compaction		Tree shading on borders	Fire damage Hail damage
Old fence lines		Prior streams	Crop lodging
Harvesting issues			

Adapted from L Lotz (1997) Yield monitors and maps: making decisions. *Ohio State University Fact Sheet AEX-550-97*, Food, Agricultural and Biological Engineering. http://ohioline.osu.edu/aex-fact/0550.html.

carry through subsequent years. Contour banks or abrupt changes in landscape usually affect growth conditions and are often noticeable as artificially low-yielding nonlinear features (b). Any broad differences in yield-affecting management practices will not usually remove the natural, irregular yield pattern but will raise or lower the overall amount of yield. In this example, the winter wheat crop was sown after the majority of the field had a summer fallow period. The northern section (c) had grown a sorghum crop during the summer, so the area had access to much less stored soil moisture than the rest of the field during the winter season.

Combining yield and protein to identify major limiting factors

Nitrogen agronomy plays a crucial role in the production of a profitable grain crop. Nitrogen contributes to the final yield of each plant through its multiple roles in general plant health and its specific roles in energy generation through chlorophyll. At the end of the growth cycle, grain crop plants move nitrogen to the seed to be stored as protein. At harvest, the protein content of the seed can be used to indicate the adequacy of nitrogen availability and uptake during the growing season. Table 6.2 documents these critical protein values for wheat, barley and sorghum crops.

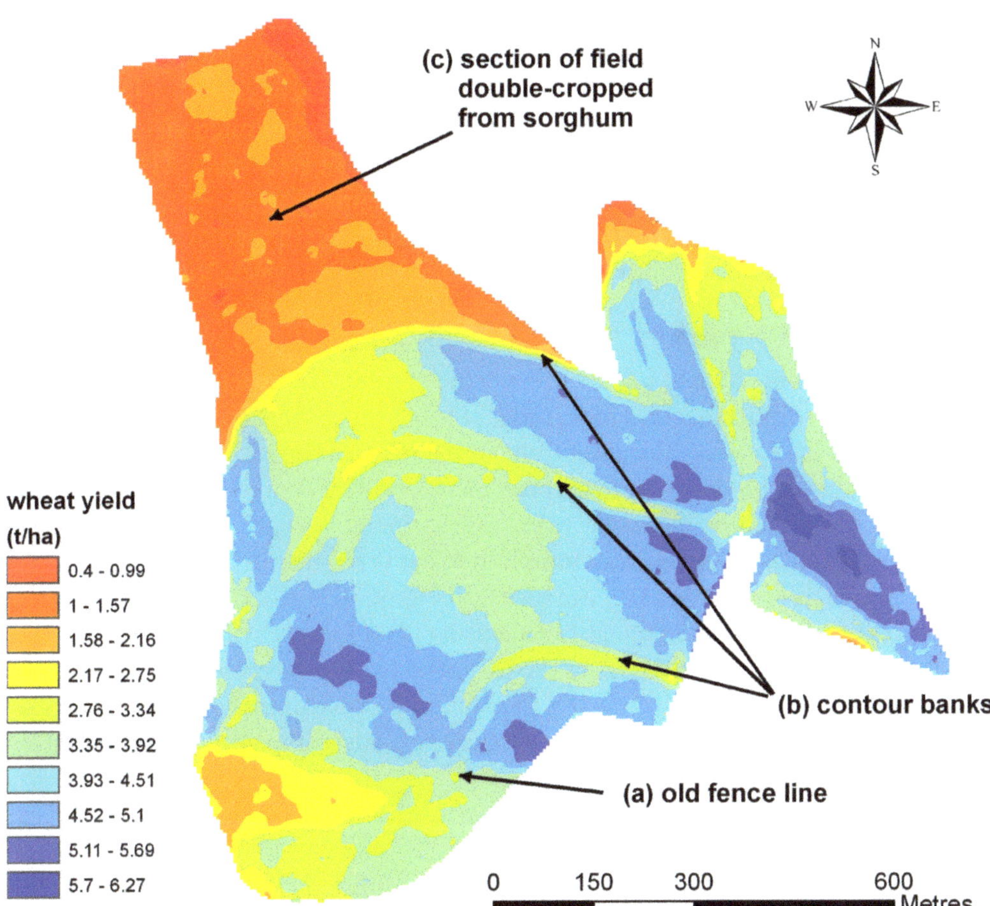

Figure 6.19: A yield map showing management-induced anomalies. The effects on yield of (a) an old fence line, (b) contour banks and (c) different crop rotation management.

Table 6.2: Grain protein thresholds for crops, which indicate the levels of soil nitrogen available to the crop during the season

Indicated nitrogen (N) supply	Wheat grain protein content (%)	Barley grain protein content (%)	Sorghum grain protein content (%)
Acute N deficiency. Grain yield would have increased with increased N supply.	<11.5	<11	<9
Marginal N deficiency. Grain yield will increase and protein will increase with increasing N supply.	11.5–12.5	11–12	9–10
N not limiting yield. Higher N supply may increase grain protein.	>12.5	>12	>10

From MJ Cahill & WM Strong (1996) Are district nitrogen recipes obsolete? In *Proceedings of the 8th Australian Agronomy Conference.* (Eds DL Michalk and JE Pratley) pp. 108–111. Australian Society of Agronomy.

Wheat grain with a protein content greater than 12.5% is indicative of an adequate supply of nitrogen being available to achieve a seasonally regulated maximum yield. However, if grain protein content has reached this level but yield is below expected, it is likely that a lack of soil moisture has limited yield. If the expected yield goal has been reached but protein content is less than 11%, nitrogen supplies can be considered to be limiting to yield. Figure 6.20 shows how this information can be used with a crop yield and grain protein map to identify areas where nitrogen has been limiting, and so enable targeted investigation.

Interpreting soil EC_a maps

The measured value of EC_a at a site reflects the combined impacts of a number of agronomically important soil properties. Table 6.3 shows how changes in these soil properties generally affect the magnitude of the EC_a readings.

Because these soil properties are combined to varying degrees across the landscape, it is presently impossible to provide a general interpretation guide based on EC_a levels alone. However, local sampling of the variability in these soil properties at different levels of recorded EC_a can be used to 'ground-truth' the data. Maps of the soil properties at the site can then be made using these local relationships.

While the major contributors to the EC_a need to be determined at each site, the relative patterns of variability usually remain quite stable over the long term, even though the actual EC_a levels will vary between seasons as moisture and nutrient levels change.

Interpreting soil gamma-radiometric maps

The measured value of the total gamma radiation emitted from the soil at a site reflects the combined impacts of a number of agronomically important soil properties. Table 6.4 shows how changes in these soil properties generally affect the magnitude of the total gamma radiation count recorded by a gamma radiometer.

As for the EC_a response, the combination of soil properties in varying degrees across the landscape makes it impossible to provide a general interpretation guide

Table 6.3: General impact of influential soil properties on EC_a (CEC = cation exchange capacity)

	Field properties						
	Material	**Soil texture**	**Clay type**	**Moisture potential**	**CEC**	**Salinity**	**Nutrients**
Higher EC_a ↑	Soil	Clay	Smectite	Saturated	High CEC	High EC	High N, P, K, Ca, Mg, Na
	Stony soil	Silt	Illite	Field capacity			
Lower EC_a	Rock	Sand	Kaolin	Wilting point	Low CEC	Low EC	Low N, P, K, Ca, Mg, Na

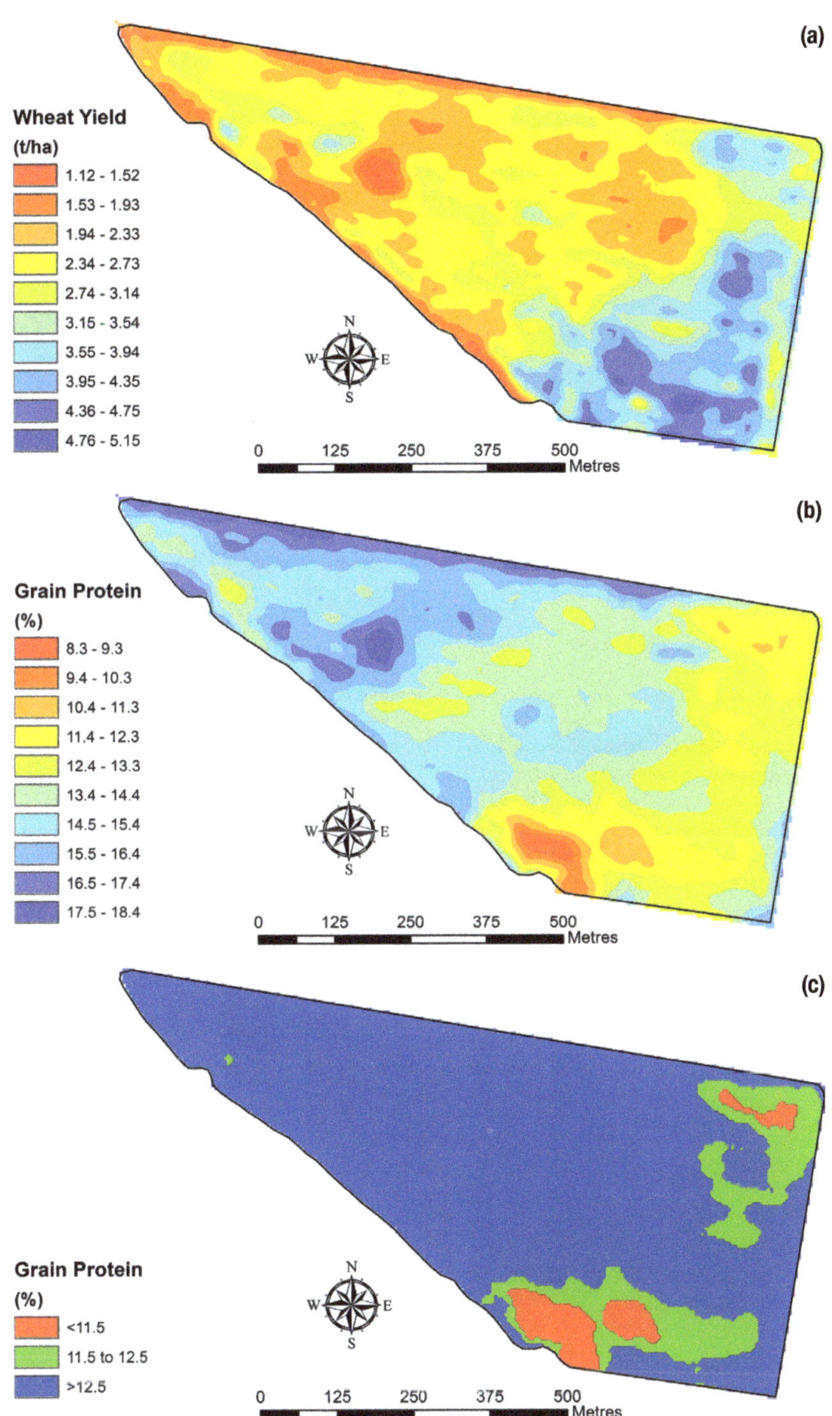

Figure 6.20: (a) Wheat yield map, (b) wheat grain protein map and (c) a combination of the two maps, identifying blue areas where the field has received adequate nitrogen for the seasonal yield, green areas where the supply of nitrogen has been marginal and red areas where nitrogen supply has been inadequate.

Table 6.4: General impact of influential soil properties on total gamma count (TC) of soil gamma emissions

Field properties							
	Material	**Soil texture**	**Clay type**	**Moisture potential**	**CEC**	**Depth to rock or gravel**	**Nutrients**
Higher TC ↑	Rock	Clay	Smectite	Wilting point	High CEC	Shallow	High K
	Stony soil	Silt	Illite	Field capacity			
Lower TC	Soil	Sand	Kaolin	Saturated	Low CEC	Deep	Low K

based on gamma-radiometric levels alone. However, local sampling of the variability in these soil properties at different levels of recorded emissions can be used to ground-truth the data. Maps of the soil properties at the site can then be made using these local relationships.

While the major contributors to the gamma emissions from the soil need to be determined at each site, the relative patterns of variability usually remain quite stable over the long term.

Using both soil EC_a and gamma-radiometric maps for interpretation

As can be seen from Tables 6.3 and 6.4, the two types of soil measurement are essentially reversed in their response to rock or stones, soil moisture content and the salinity status of the soil. The potassium band of the gamma radiometer is specifically influenced by changes in potassium in the soil. These differences mean that a comparison of maps from an EC_a and a gamma-radiometric survey of the same site can prove very useful in pinpointing areas of soil with gravel near the surface, especially in generally sandy soils. The two maps can also be used to diagnose areas where soil moisture or salinity (and not necessarily clay or CEC) are likely to be influencing high EC_a readings.

In Figure 6.21, the EC_a survey (a) and the thorium emissions of a gamma-radiometric survey (b) both show a pattern of variability across a farm in Western Australia. While there are some similarities in the patterns, there are areas where both the EC_a and gamma emissions are low (site 1) and where the EC_a is low but the gamma emissions are high (site 2). The difference in the gamma emissions is due to the presence of a 'uniform' sand at site 1 (Figure 6.22a) and a very shallow sand over gravel at site 2 (Figure 6.22b). The gravel (rock) is as poorly conductive (low EC_a) as the sand, but it has a high gamma emission. The areas of high EC_a at site 3 in Figure 6.21a are not picked up in the gamma survey. These higher measurements are usually due to an increase in soil moisture content and an associated rise in EC_a caused by salinity in these types of soil.

In Figure 6.23, the relationship between the yield map and the individual soil maps is not strong. However, when the two soil maps are considered together, the

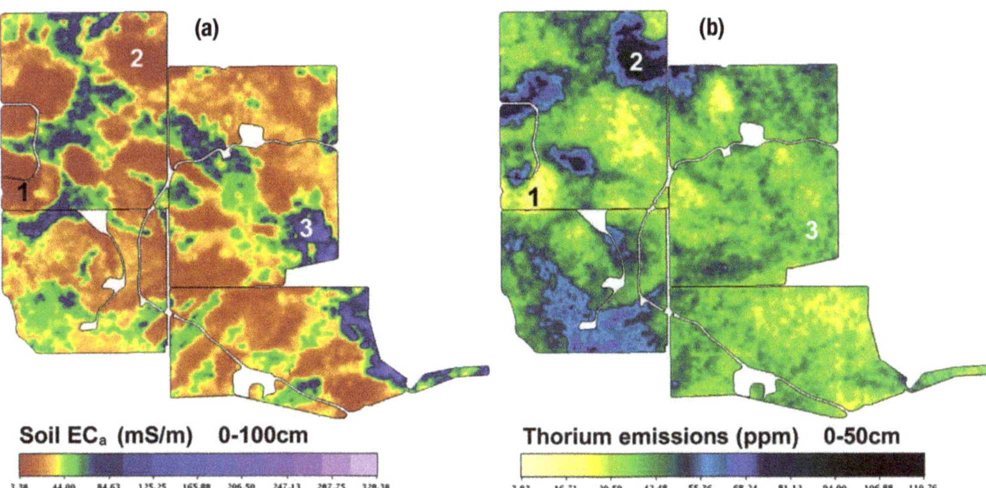

Figure 6.21: Results of (a) an EC_a survey and (b) the thorium band from a gamma-radiometric survey on a farm in Western Australia. (Images provided by Precision Agronomics Australia)

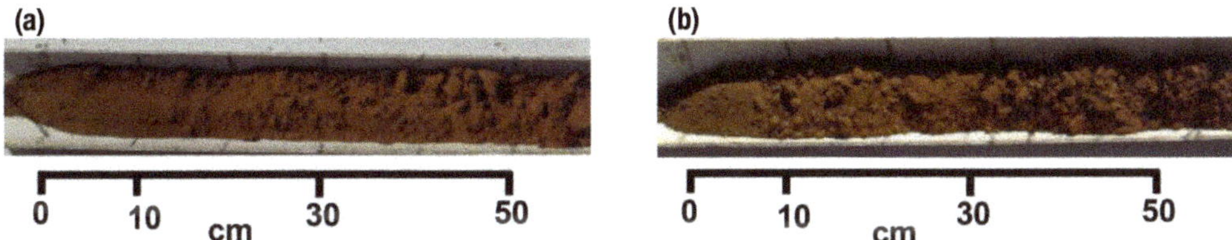

Figure 6.22: Soil cores taken from (a) site 1 and (b) site 2 in Figure 6.19, showing the marked increase in amount of gravel near the surface in (b). (Images provided by Precision Agronomics Australia)

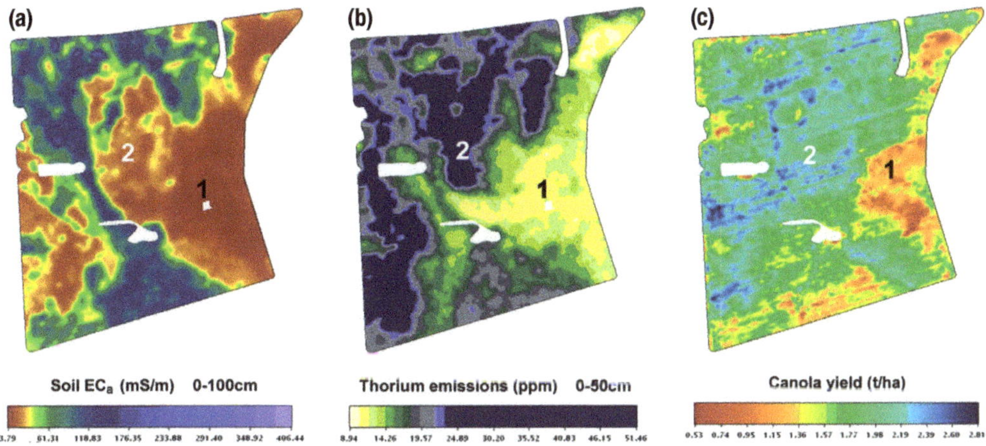

Figure 6.23: Maps of (a) EC_a, (b) thorium gamma emissions and (c) canola yield from a field in Western Australia. (Images provided by Precision Agronomics Australia)

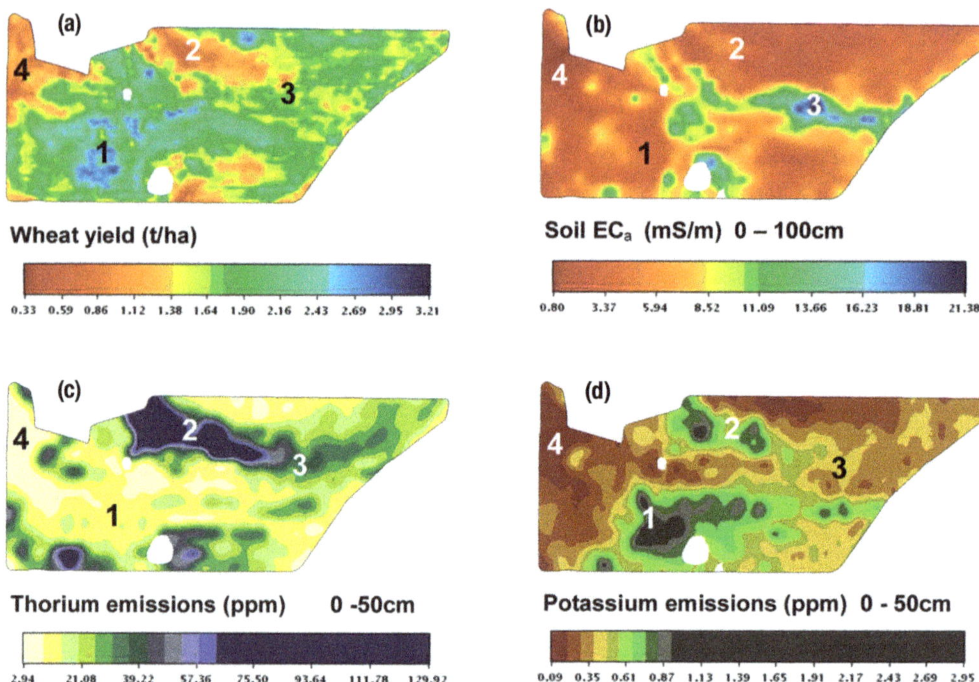

Figure 6.24: Maps of (a) wheat yield, (b) soil EC_a, (c) thorium gamma emissions and (d) canola yield from a field in Western Australia. (Images provided by Precision Agronomics Australia)

interpretation of the yield variability improves. The gamma radiometrics again helps to pick the difference between the uniform sand at site 1 and the shallower sand over gravel in site 2. The canola yield map (Figure 6.23c) shows that at site 1, where there is the least gravel, the yield is lower than site 2. Using only the EC_a map to try to understand the effect of soil variability on the canola yield would be difficult in this case.

Figure 6.24 shows a field where approximately 90% of the area is below an EC_a reading of 10 mS/m, indicating a sandy soil. With the addition of the gamma radiometrics, a distinction between areas with gravel inclusions (site 2) and uniform sands (site 1) can be made. In this case, the soil with more gravel closer to the surface reduces the yield of the wheat crop. At site 3, where the EC_a is the highest in the field, the thorium emissions also remain reasonably high but the yield increases to the field average. The interpretation is that there is a potential increase in soil moisture at site 3.

This example also highlights the need to consider other layers of information when interpretation is not straightforward. At site 4, both the EC_a and thorium emissions are as low as site 1, but the wheat yield is low. The potassium emissions (Figure 6.24d) show that the potassium levels are potentially lower at site 4 than site 1, but other influences, such as terrain, other soil constraints and pest pressures should always be considered during interpretation at a site.

Key points

- Precision agriculture (PA) data that is collected to map variation in aspects of farm production can be either 'point' or 'raster' data. Point data consists of individual measurements with their own location information attached (e.g. yield monitor data). Raster data has measurements allocated to a single cell in a grid of cells covering an area of interest (e.g. satellite imagery).
- It is important to put as much effort as possible into gathering good-quality data and to use available software tools to filter out any poor-quality data before making a map.
- There are a number of ways to make maps from raw PA data. The type of data and the method with which it was collected should determine the mapping method used.
- It is important to understand and use the full capabilities of a software mapping package. Do not just settle for default settings every time.
- Highly accurate point data can be displayed as point or grid maps if the number of points is large.
- Surface maps are best for the majority of point data gathered in PA operations. They make point data easier to interpret and allow various analyses of data gathered by different sensors or at different times.
- The process used to make a map affects the appearance of a map and any future analysis.
- Choose an appropriate method for allocating data to colour ranges in a map, because this affects the interpretation.
- To compare maps over time, it is necessary to use a consistent allocation methodology. Different methods tend to enhance different sections of the data.
- For the clearest interpretation and comparison of crop yield maps it is advisable to use the equal interval allocation method.
- All good PA maps have a legend that explains the meaning of the map symbols, colours or shades. It also displays the physical scale and the orientation of the mapped area.
- Spending time visually checking a map for anomalies will reduce the possibility of making an incorrect assessment of management requirements.
- Natural variation is usually seen as an irregular pattern with gradual changes between different levels.
- Linear patterns reflect management operations (e.g. old fence lines, sowing/fertiliser/agrochemical application problems and machinery errors) or the effect of underlying infrastructure (e.g. utility easements).
- Irregular patterns that display abrupt changes are often the result of management operations (e.g. chemical spray drift) or sharp topography changes.
- Soil apparent electrical conductivity (EC_a) and gamma emission maps can be useful in interpreting the effects of soil variability on yield. A combination of the two techniques can prove even more useful in diagnosing changes in soil properties, especially in the sandy soils of Western Australia.

7

Yield variability and site-specific crop management

There is a large list of important components of a farming operation for which it might be useful to have data on the extent of variability. For some components, such as fertiliser quality, farmers rely on outside companies to minimise the variation and so remove the need for substantial on-farm management. Others, such as soil properties, pest and disease outbreaks and crop yield will vary on each farm. Local knowledge about variability in these parts of the farming system can be used to build site-specific crop management (SSCM) strategies. SSCM can be used to identify and treat any areas where yield potential can be improved or to better match input use to the natural variation in yield potential across a field or farm.

Understanding variability

There are two main types of variability to consider when exploring how PA may be useful for a farming operation.

Spatial variability

The variability in soil, crop, landscape and environmental attributes that occurs across a certain area.

Temporal variability

The variability in soil, crop and environmental attributes that occurs within a certain area at different measurement times.

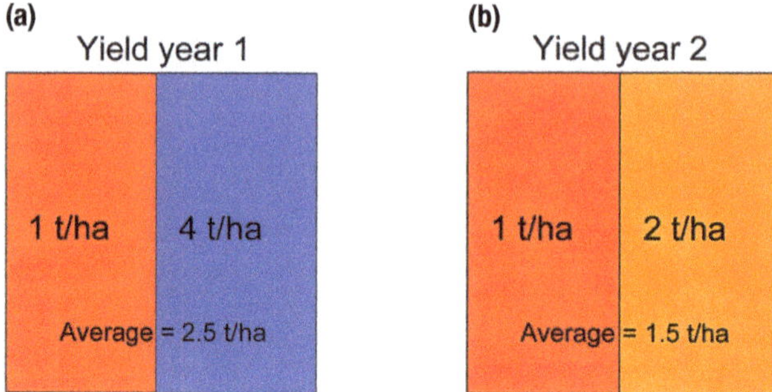

Figure 7.1: An example of crop yield spatial and temporary variability. (a) Year 1 average crop yield = 2.5 t/ha. (b) Year 2 average crop yield = 1.5 t/ha.

The concept of spatial variation is demonstrated in Figure 7.1(a) where the crop yield from a hypothetical field averages 2.5 t/ha but comes from two equal-size parts of the field that yield 1 t/ha and 4 t/ha respectively. Temporal variability can be seen when comparing Figure 7.1(a) and 7.1(b). In the same field, the average crop yield for two successive seasons changes from 2.5 t/ha to 1.5 t/ha.

While these two concepts of variability are at the heart of PA, it is the amount and pattern of variation that are important for decision-making.

Again, when comparing Figures 7.1(a) and 7.1(b), the same pattern of variability is evident but the difference between the areas is much less in Figure 7.1(b). The potential benefit from tailoring different management to the two different areas in the field is less in Figure 7.1(b). In Figures 7.2(a) and 7.2(b), the average yield (2.5 t/ha) and the amount of variation (50% = 1 t/ha; 50% = 4 t/ha) are the same. What is different is the pattern or distribution of the variation. From a practical viewpoint, tailoring different management to the different yield potentials in the field would be much easier in Figure 7.2(a).

The area over which these measures of variability may be made can range from within-field to whole-farm scale. For measuring temporal variability, the time

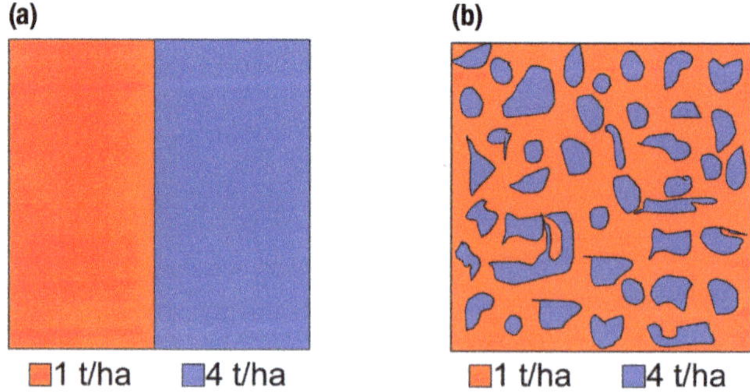

Figure 7.2: An example of differences in the pattern of spatial variability. The average yield (2.5 t/ha) and the amount of variation (50% = 1 t/ha; 50% = 4 t/ha) is the same in both (a) and (b).

between measurements is typically from one significant crop growth stage to another or from one season to another.

Impact of variation in soil attributes on crop yield variability

Soil texture

Soil texture is usually measured by the percentages of sand, silt, clay and gravel in a soil. It is possible to estimate the soil texture in the field using simple hand-texturing techniques. The greater the quantity of sand-sized particles in a soil, the coarser is the soil texture. Soil texture influences the crop yield potential of a site by contributing to the variation in nutrient storage and availability, water-holding and transport, the binding/breakdown of agrochemicals and soil stability to disruptive forces. The general influences of different soil textures on these secondary soil properties is summarised in Table 7.1.

Spatial variation in soil texture within a field may contribute to the final pattern of crop yield. Soil EC_a and gamma-radiometric surveys have been used to locate major changes in soil texture and can be used to select soil sampling sites to characterise these changes.

Table 7.1: General physico-chemical properties of soil in different texture classes. 1 = very little, 3 = medium, 5 = large

	Soil texture classes (broadly defined by average % clay content)				
Property	Sands (<5%)	Sandy loams (15%)	Loams (20%)	Clay loams (30%)	Clays (>50%)
Total plant available water (PAWC)	1	2–3	4–5	3–4	3
Rate of water infiltration	5	3–4	3	2–3	2
Likelihood of hard setting, crusting or sealing	2	4	3–4	3	2–3
Susceptibility to water erosion	2	2–3	4–5	3	2–3
Susceptibility to wind erosion	3–4	4	4–5	3	1–2
Capacity to shrink/swell	1	1–2	2	2–3	4–5
Time until trafficable after rain or irrigation	2	3	3	3	4–5
Susceptibility to compaction	2	3	3–4	2	4
Cation exchange capacity (CEC)	2	2–3	3	3–4	4–5
pH buffering capacity	2	2–3	3	3–4	4–5
Potential for leaching of nutrients and herbicides	4–5	3–4	3	2–3	2

Soil structure

Soil structure may be simply described as the arrangement of particles that form the soil and the distribution of spaces between these solid particles. The structure of the soil affects the physical penetration, growth and anchorage of roots. It also regulates the air/moisture balance (which is important for good plant growth and microbial activity) as well as the soil drainage/water-holding characteristics and the erosion potential.

The structure of a soil can be degraded by:

- mechanical forces such as compaction by vehicles, implements or animals;
- disruption by cultivation;
- wetting and drying cycles in the soil;
- loss of organic matter;
- increase in sodicity.

A decline in soil structure may result in a broad range of adverse effects on crop growth. These are commonly:

- reduction in the availability of oxygen required for metabolic processes;
- reduced infiltration, increased ponding, run-off and loss by evaporation;
- reduction in the plant available water capacity (PAWC);
- restriction in root volume;
- reduced root elongation.

Structural degradation can influence yield but the degree of impact may be determined by the season. For example, in a wet year the effects of compaction may cause increased waterlogging, while in a dry year, especially in a hard-setting soil, compaction effects will reduce the quantity of soil water and may restrict root exploration. The different impacts of soil structure decline mean that, when interpreting crop variability in a field, it is important to consider variability over several years with the recorded seasonal conditions.

Any spatial variability in soil structure can affect the efficient use of inputs, such as fuel and nutrients, and contribute to the spatial pattern of crop yield within a field. Options for detecting the spatial variability in soil structure include measurements of soil strength using tillage draught, tractor fuel consumption during tillage and cone-penetrometer resistance; pore/solid relationships via air permeametry and soil bulk density sampling; compaction layer location using ground penetrating radar or EC_a surveys.

Soil pits can be used in combination with spatial information to observe soil strength and plant root interaction. For identifying spatial variability in surface crusting, susceptible areas are probably best detected by sampling for the underlying cause (e.g. sodicity or low organic matter) or by visual assessment after the soil dries from a rainfall event that was sufficient to cause crusting.

Soil depth

Soil depth is defined here as the depth of soil that can be readily accessed by crop roots. This depth is a very important parameter because it greatly influences the total quantity of water and nutrients available to the plant. It may be the depth of soil to bedrock or to a subsoil layer that is impenetrable by plant roots. Some subsoil layers are a permanent feature of the soil resulting from the way the soil was formed (e.g. layers high in clay, boron, salinity, acidity or alkalinity). Others (e.g. a plough pan) may be caused by previous management. While the actual rooting depth of a plant depends on the species, it is also controlled by the amount and location of soil water. If there is no soil water at depth, roots will tend to grow only in the surface soil. It should be noted that any water or nutrients that pass below the actual rooting depth of a plant are unavailable and may contribute to environmental damage, such as salinity or pollution.

In some shallow soils, the spatial variation in depth to bedrock may be determined by using a graduated push probe. This operation is most easily carried out when the soil is wet. If there is no information on the variation in depth, probing over a coarse sampling grid will indicate the magnitude of the variation and areas where probing on a finer grid would be worthwhile. There is no point probing on a fine grid in an area where there is little variation.

Probing may also be used to locate some restricting subsoil layers when there is a sufficiently large difference in strength between the problem layer and the overlying soil. Some relatively shallow and thin layers (e.g. a plough pan) may be difficult to locate when the soil is wet and the strength of all layers is low. When there are multiple layers, it may be difficult to differentiate between them using a push probe, especially if the depth and thickness of the layers are highly variable.

In these circumstances, coring or other sampling will be needed to identity the layers. Where there is a marked difference in the properties of one layer and another deeper layer, an EC_a and/or gamma-radiometric survey may help to indicate the spatial extent of the changes in depth of the surface layer once it is calibrated with probing or sampling.

Soil organic matter

In most soil types the amount of soil organic matter (OM) provides an indicator of the inherent soil fertility. OM plays a significant role in stabilising soil structure by binding soil particles together and by storing and releasing nutrients. The amount and type of OM also influence the quality and quantity of soil microbial activity and the provision of mineralisable nitrogen, phosphorus and sulphur. The importance of OM in storage and release of moisture and plant available nutrients increases as the percentage clay content decreases. The amount of OM present also affects the degree and manner in which agrochemicals are adsorbed or broken down. The more OM, the greater the degree of binding that may occur.

Spatial variation in OM may therefore affect decisions about fertiliser and agrochemical application rates, along with carbon retention programs. Spatial variability in OM in the surface soil can be measured using spectrometers.

Soil water

In most regions of Australia, a lack of soil water is the dominant yield-limiting factor. Studies have shown that 50% of the variation in wheat yield can often be explained by variability in soil moisture at the time of sowing. Variability in soil water content also influences soil biological activity and soil temperature, which in turn affect nutrient uptake in roots and root elongation.

The total amount of water that can be stored in the soil profile and made available to plants is called the soil's PAWC. The PAWC is essentially governed by the soil's structure, texture and depth. Generally:

- good structure provides more pore area for water to be stored;
- clay holds more water than sand but it also holds it more tightly than sand as the soil dries;
- a deeper soil provides more volume to store water.

Although the soil texture or depth cannot easily be altered, soil structure and crop rooting depth can in some situations be changed by management practices. Removing subsoil constraints such as compaction, salinity, boron, sodicity and layers of extreme pH can help roots extract all the available water at depth.

Spatial variation in PAWC is important in regions where cropping relies heavily on stored soil water. It is less important where in-season rainfall is sufficiently uniform to supply crop needs during the season. However, the factors that contribute to a larger PAWC also allow the soil to hold more water between rain events.

While the PAWC describes the overall moisture-holding ability of the soil, the plant available water (PAW) is the total amount of water stored in the soil profile and available for crop uptake at a particular time (e.g. at sowing). Although a lack of soil water is usually a limiting factor to crop yield, excess water can also be important. Waterlogging can be very detrimental to crop growth and yield, even if it is only temporary.

Variability in the amount of soil moisture stored and made available around a field may significantly affect the observed spatial pattern of crop growth and yield. Measuring the spatial variability in PAW at a given time can be performed using a range of measurement technologies in different parts of a field. Soil EC_a surveys have been tested as a potential low-cost method of collecting this data just before sowing. Any technique needs to be calibrated by soil type as the instruments are all influenced by differences in soil physical properties. Data has shown that as soil water content increases, its spatial variability decreases. This maxim forms part of the rationale for crop irrigation.

Soil pH

Soil pH is a measure of the acidity or alkalinity of the soil. An acid soil has a pH value less than 7, a neutral soil has a pH of 7 and an alkaline soil has a pH larger than 7. The pH of the soil is important because:

- most agronomic crops and soil organisms prefer a neutral to slightly acidic soil;
- some soil-borne diseases tend to prefer a particular pH range;
- pH levels can affect the availability of nutrients in the soil.

Spatial variation of greater than 1 pH unit across a field will induce variability in the plant availability of nutrients, even if fertiliser is applied in uniform quantities. At low pH levels (a high level of H^+ ions) aluminium and manganese become highly available, often at levels toxic to plants.

Measurements of spatial variability in topsoil pH can be made using on-the-go sensing systems, but ideally a measurement of changes in the buffering capacity (BC) of the soil could be made. The BC is a measure of the soil's ability to resist changes in pH and is controlled by the geology, soil texture and soil OM content. It is used in calculations to determine the amount of lime required to increase the soil pH by one unit (e.g. from pH 6 to pH 7). A measure of the spatial variability in BC would therefore be useful to prescribe variable-rate applications of lime.

Soil nutrients

The sufficient supply of macro- and micronutrients is crucial to growing crops. Variability in soil nutrients is ultimately governed by variability in soil physical factors and moisture cycles, along with the soil pH. These factors influence both plant root growth and the supply of nutrients to the roots by controlling the total quantity of available nutrient, the rate of movement and the ease of delivery to the roots.

The spatial variability in the major nutrients is usually related to increasing nutrient mobility in the soil (nitrogen is more mobile than potassium, which is more mobile than phosphorus). Adding fertiliser and increasing OM increases the total amount of, and the variability in, nutrient content in the soil. The effect of variability in soil nutrient concentrations on crop yield is magnified by the fact that substantially less than 50% of a crop root system is able to absorb nutrients.

Measurement of the spatial variability in soil nutrients has focused on determining changes in the levels of nitrogen, phosphorus and potassium prior to sowing and the level of deep soil nitrogen prior to in-crop nitrogen applications. The levels of soil nutrients are generally determined by taking soil samples, which are then analysed in a laboratory. In-field soil nutrient sensing systems are a development goal of many companies and research institutions.

Impact of variation in other important factors on crop variability

Weeds

The spatial variation of weed plants in a field is a function of the species (weed and crop), environmental conditions and previous and current cultural practices. In general, weeds develop in patches within a field and the presence of weed species is different between fields. However, the pattern within a field usually shows great similarity between years. The difference in weed development between fields means that individual field recommendations for treatment will be required. Patchiness means that parts of a field may remain weed-free and therefore require no treatment.

In the context of PA, the spatial distribution of weeds is measured using a number of different tactics. The location of weed patches can be identified using GNSS equipment and mapped using hand-held devices, mapped during harvest operations or mapped via remote sensing. Real-time weed-sensing systems can identify weed infestations in-crop.

Insects

Insect pest damage also appears in patches but its appearance is often more dynamic than that of weeds. The spatial variation in insect damage is a function of the insect species along with the insect and host crop life-cycle stage. It can also be linked to the impact of spatial variation in soil and host plant health/vigour on insect immigration, colonisation, reproduction, emigration, predation and mortality. Many insect pests are attracted to healthy or better-growing crop plants because they provide the greatest chance of sustaining their life-cycle. Obviously many insect pests are crop-specific, but climatic conditions along with spatial variation in soil conditions may interact to provide areas in a field that are more attractive for attack. Therefore, knowledge of the spatial variability in soil and cropping conditions before pest infection might aid in predicting the spatial distribution of subsequent insect infestations.

Diseases

The risk of disease developing in crops is dependent on many local and regional factors. Soil- and stubble-borne disease severity is mostly determined by the carryover of the pathogen at the site. For example, yield loss caused by cereal cyst nematode is determined by the number of eggs in the soil, and for take-all it is the number of infected crowns and roots persisting from the previous year. Therefore, it is possible to estimate the risk of such diseases by testing soil samples, especially as DNA-based tests are available.

Leaf diseases, such as rusts, often result from spores blown from neighbouring farms or other districts. In such cases, disease prediction needs to consider factors at a district or regional scale, such as climatic conditions.

Australian research has shown that areas of different yield potential within a field can have different disease risks which may be managed using differential agrochemical application. This practice is more suited to soil- and stubble-borne diseases. Inter-row sowing has also been effective in minimising disease infection in the important early stages of crop development.

The targeted application of late fungicide treatments to protect high-yielding areas may also be useful in combination with early sprays applied across the whole field to delay the build-up of leaf disease epidemics.

Terrain

Elevation and slope changes can be an important cause of crop yield variability because:

- they are a strong influence in the soil-forming process. Within a field, soil at higher elevation is usually shallower and of coarser texture than soil at lower elevation;
- areas of relatively lower elevation can collect cold air which may lead to increased frost susceptibility;
- these changes can control the way water accumulates or moves across and within the soil.

The direction in which areas of sloping land face, known as the aspect, can have direct effects on yield variability. In the southern hemisphere, north- and west-facing slopes are hotter and dry more quickly than those facing south and east. Depending on the direction of the predominant weather systems in a region, there may be differences in rainfall due to aspect.

Past crop management

The crops and associated management operations applied in past seasons can have a large influence on current season yield variation. For example:

- variability in a legume crop grown over a field will contribute to variability in nitrogen reserves for the next season;
- the residual effect of crop-specific herbicides may be variable, depending on changes in soil type;
- faulty fertiliser or agrochemical application may have effects for several seasons.

It is important for growers to record and consider past management operations when PA data layers are being interpreted.

All the factors described above should be considered when investigating the causes of yield variability within a field and across a farm.

Applying SSCM

SSCM aims to use knowledge of the importance of soil properties and other factors in crop production and yield potential to better identify and understand the reasons for changes in growth and yield within a field. The goal is to improve decision-making about the use of inputs such as fertilisers, agrochemicals, fungicides, lime, gypsum, water and fuel, to better match spatial changes in the requirements of the soil and crop. A better match should mean that inputs (including soil moisture) are used efficiently, profits are maximised and waste is minimised.

The way that soil properties, pest and disease loads, terrain and past management interact and vary across each field on each farm is unique. This type of variability is 'site-specific' and is the reason why no single SSCM prescription can be shared between farms. Because of site-specific variation, the best information for determining SSCM options for each farm or field will undoubtedly come from within its own boundaries.

How to begin

It is important to sort out the best 'average' management options for fields and crops before considering the need and requirements for variably applying inputs.

- Any large-scale soil pH and sodicity or salinity issues should be identified. Soil samples, taken from across the farm, should be analysed to correctly identify any required treatment.
- The effectiveness of weed control strategies should be examined and improved if required.
- Elevation mapping using GNSS will allow field boundaries and working directions to be altered to improve water management if necessary.
- Average yield information gathered for each crop should be used to re-examine crop yield goals on the farm and to ensure average nutrient application is suitable. Yield monitors can help, especially if historical records are missing or the concurrent use of off-farm delivery and on-farm storage makes it difficult to link tonnages to specific fields.
- Vehicle navigation aids (guidance or autosteer) offer immediate benefits through reducing the overlap during fertiliser, sowing and chemical operations and providing improvements in soil structure. Autosteer systems can be used to sow into the inter-row between last year's stubble to reduce the incidence of root diseases.

Once any alterations are put in place to optimise the average or uniform agronomy on a farm, the aim should be to test whether SSCM can be used to improve management decisions on the farm. This requires that within an area:

- the amount and pattern of variability in yield and soil conditions can be measured;
- reasons for the variability can be determined;
- the agronomic implications of the variability and its causes present a practical opportunity to vary management.

The easiest way to start is to investigate areas of high, average and low yield within fields. A yield monitor makes it easy to identify these areas and provide the locations to then perform:

- a complete soil profile examination and analysis;
- crop observation for disease or pests;
- assessment of the position in the landscape (slope, hill, hollow).

This information should be used, in conjunction with local agronomic understanding and advice, to determine whether the amount, pattern and causes of variability warrant changing from a uniform application of inputs. With the understanding that local variability should be the driving factor, it is possible to provide a general strategy for the incorporation of SSCM on any farm (Figure 7.3).

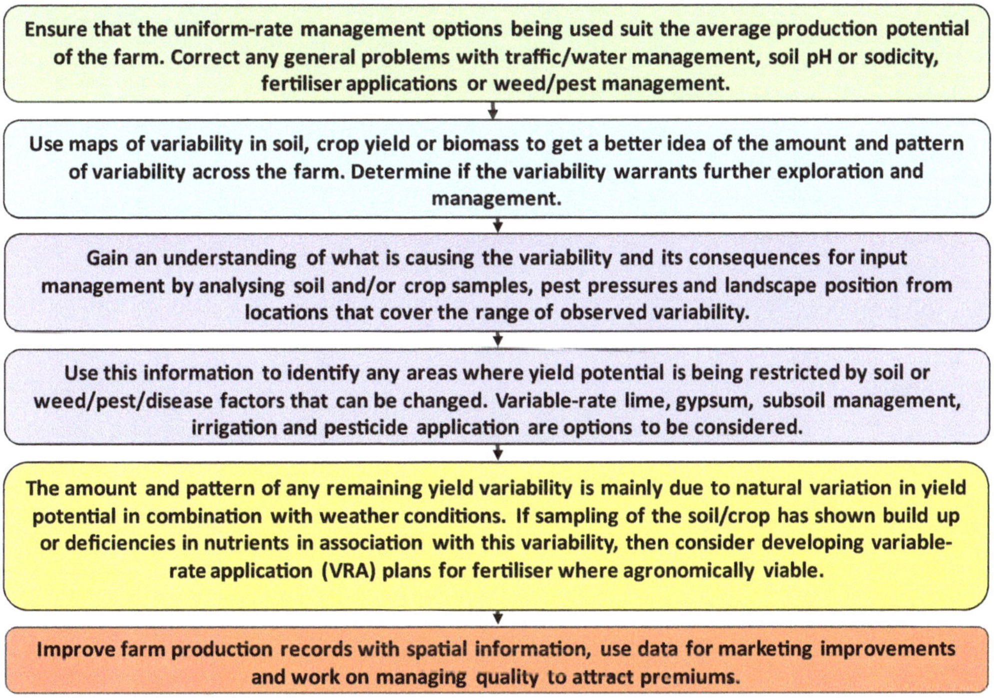

Figure 7.3: General strategy to drive the incorporation of SSCM on a farm.

Variable-rate application of inputs

Responding in a practical way to variation in production potential requires that input rates be adjusted to match changing local requirements. Variable-rate application (VRA) can be aimed at any rate-based operations that influence crop yield and can be done with the help of variable-rate technology (VRT) or by using manual switching or multi-pass applications. Using VRT makes the job less stressful and allows more sophisticated positioning and rate adjustments. The main target operations for VRA in grain cropping are:

- fertiliser;
- soil ameliorants (lime, gypsum);
- agrochemicals (herbicide, insecticide, fungicide);
- sowing;
- tillage;
- irrigation.

Ideally, the use of VRA should optimise both the economic and environmental outcomes of a field. In Australia, the economic considerations dominate as there is little regulatory auditing of the levels of approved chemicals in the agricultural environment. However, most crop producers are well aware that maintaining a healthy environment is important for sustainability and economic success.

The potential benefits of VRA are generally higher when:

- the amount of spatial variation is larger;
- the pattern of spatial variability tends more towards coherent patches. This usually means fewer rate changes are required;
- the pattern of variability is driven by spatial rather than temporal factors, so it is likely to be relatively stable from season to season and easier to formulate VRA plans;
- the unit cost of input is high relative to the price paid for the crop.

VRA options

In general, VRA can be used to deal with spatial variability at three different scales:

- whole-field variability;
- variability between potential management classes;
- variability within potential management classes.

Whole-field variability

Input rate changes are made across a continuous range to deal with continuous variability in crop requirements. It is a prerequisite that the variability in requirements can be determined accurately at a fine scale across a field.

This type of operation may utilise map-based or real-time VRT. In map-based applications, the prescription map may be derived from a single attribute map or the combination of a number of maps. Single maps of variability can be used directly to formulate management plans when the variability and its impact on management operations are obvious and directly linked. There are a number of current options:

- topography maps can be used to realign working directions, field boundaries or seedbed formation to improve water management;
- in broadacre irrigated fields, topography maps can be used to derive cut/fill maps for levelling purposes;
- crop yield maps can be used to calculate nutrient removal and drive a VRA replacement strategy. This is most suitable for the less variable, less mobile phosphorus;
- soil pH maps made from real-time sensing systems can be used to direct lime ($CaCO_3$) application.

As more on-the-go sensing systems are commercialised for directly mapping soil properties (e.g. soil nutrients and carbon), decision-making from single maps may become more common.

At present, VRA application at the whole-field scale using combinations of maps is restricted to the standardisation and combination of a number of seasons of crop yield data to provide a continuous map of variation in the average nutrient removal.

Real-time VRT to treat continuous variability remains the domain of crop reflectance sensing systems that identify the presence and vigour/health of plants. Systems that calculate and prescribe nitrogen requirements, growth regulator and agrochemical applications have all been commercialised.

Ultimately, the technical specifications of the application equipment will control the minimum treatment area and minimum incremental rate change in any continuous VRA operation. Specifically:

- the minimum independent section width of the application equipment;
- the speed at which a rate-change can be relayed to the equipment;
- the time it takes for the equipment to change up and down rates;
- the speed of travel.

Potential management class (PMC)

In defining PMCs, fields are divided into areas (classes) that have shown differences in production potential and may require different management treatments. The variability of an attribute of interest is described by the average value in each class.

- A management class in SSCM is the total area for which a specific management treatment can be identified. Management classes are distinguished from each other based on the requirement for different management treatments.

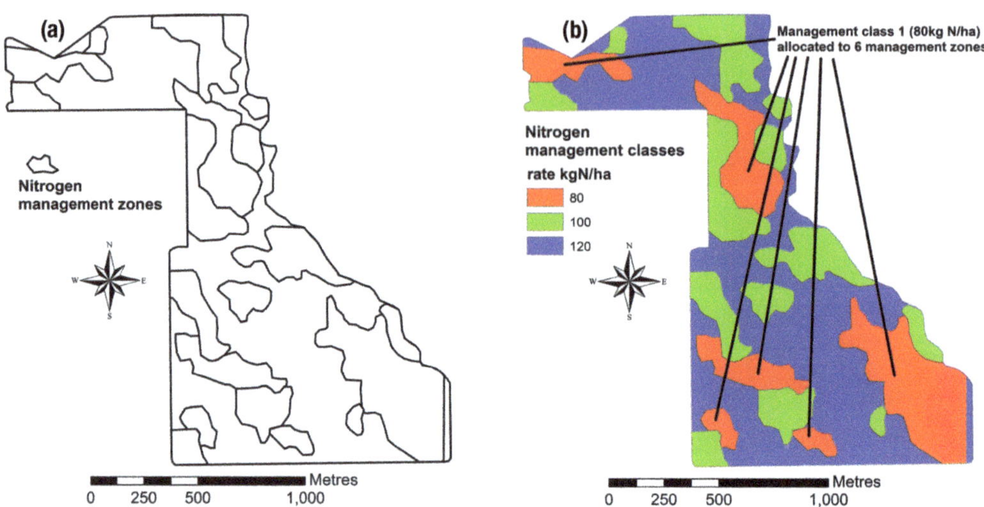

Figure 7.4: (a) The management zones for a field. (b) The management classes with nitrogen prescriptions are allocated to their respective zones.

- A management zone is an unbroken area to which a specific management class treatment is applied.
- A management class may therefore be allocated to one or more management zones within a field or farm (Figure 7.4).

Figure 7.5 shows how the PMC concept breaks the variability into a number of relative categories, with more classes increasing the ability to describe the range of variability.

This approach may be regarded as a risk-averse compromise between uniform management (single-rate application) and continuous management of whole-field variability.

Essentially, the PMCs should partition the variability within a field so that:

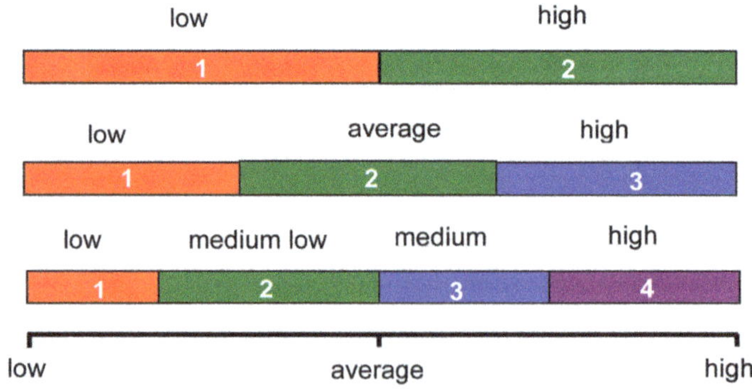

Figure 7.5: The number of management classes affects the way variability is categorised.

- within-class variability is reduced below whole-field variability;
- average within-class variability is significantly different between management classes;
- the reduction in variability will also be expressed in important attributes that have not been used to make the management classes.

The boundaries of the classes can be determined using a wide range of methods and initial information. The methods that are most widely employed include:

- growers drawing boundaries by hand using their knowledge of a field;
- drawing boundaries around different areas on yield maps or imagery;
- supervised or unsupervised classification of satellite or aerial imagery;
- cluster analysis and other statistical processes applied to one, or used to combine a number of, yield/imagery/soil/landscape maps.

Figure 7.6 shows the result of combining a number of data layers using a clustering process to form potential management classes. Using previously gathered maps of variability and statistical processes takes the guesswork out of setting boundaries, but it is important to make sure that maps correctly reflect the variability in the field and do not include problems in the patterns caused by poor past crop management or collection errors.

As can be seen from the legends in Figures 7.5 and 7.6 (e, f), when the number of management classes is increased the differences between the averages in each class decrease. As these differences decrease, the potential benefits of managing the classes differently are also reduced. Any decision on the number of management classes to use should involve consideration of maintaining an agronomically significant difference and ensuring the pattern of variation is not overly broken-up into very small unmanageable zones.

There is no set number of classes that should be identified. The decision is influenced by the treatment to be applied, the range of variability in the identifiable driver of the treatment and the importance to production of matching small incremental changes in the driver with changes in the treatment. The general practice in Australia is to use one to four management classes for a particular operation. Most commonly, one to three management classes are identified:

- one – uniform treatment;
- two – a division between high and low;
- three – high, medium and low treatments identified.

Variability within PMCs

The base rate requirements for each management class are identified prior to the application operation, but the actual input rate is modified by a measure of variability gathered during the operation within each zone/class.

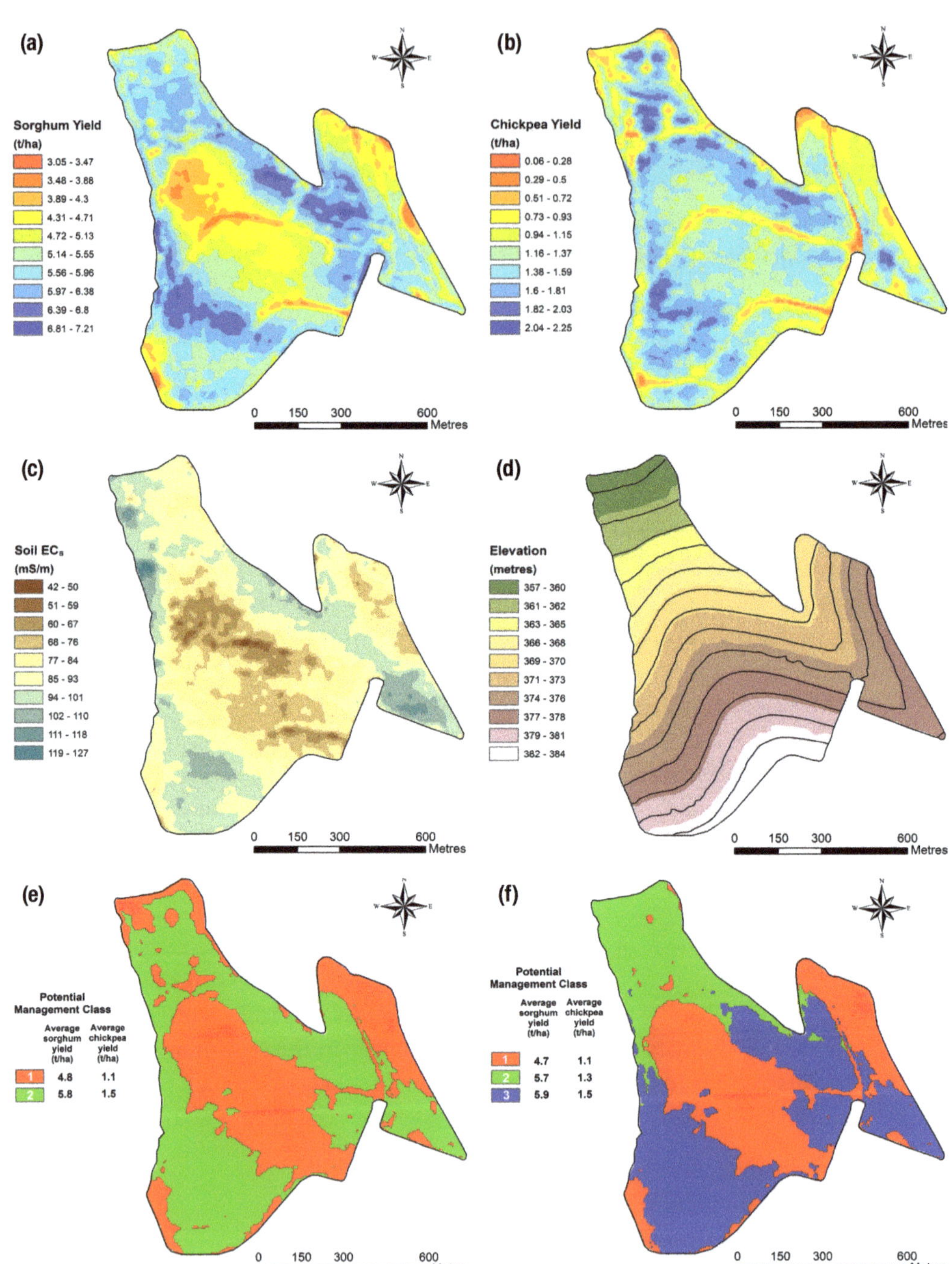

Figure 7.6: (a, b) Whole-field variability in crop yield for two seasons. (c) Soil EC_a (d) Elevation. The (e) two and (f) three potential management classes built by combining maps (a), (b), (c) and (d) in a multivariate clustering process.

This option is a mix of dealing with variability using PMCs and continuous whole-field treatment. It relies on using historical information and/or knowledge in combination with real-time VRT. Real-time systems allow the operator to set base rates; some systems also provide the option of importing base-rate maps to be combined with the data from the real-time sensor during application.

VRA operations

The four main categories of VRA operations can be defined by the scale at which information on variability is gathered and treated, and the category of VRT equipment that is used (Table 7.2). The categories are:

- map-based management class operations;
- map-based whole-field operations;
- real-time management class operations;
- real-time whole-field operations.

Table 7.2: The four categories of VRA operations are defined by the scale at which information on variability is gathered and the category of VRT equipment

	Management class treatment of variation	Whole-field treatment of variation
Map-based	A map of broad areas where different management is required is used to describe where rate changes will occur. Examples: • Nitrogen applied at three different rates to three management classes • Lime applied only in areas previously identified as requiring pH adjustment	A map of variability across a whole field is used to determine continuous rate changes. Examples: • Phosphorus applied based on the pattern of removal from previous yield map/s • Pre-emergent herbicides applied based on a map of soil organic matter content
Real-time sensors	A map of broad areas with different management requirements is used to describe where base rates will change. Real-time sensors are then used to measure variability within each management class and to vary application around the base rates. Examples: • Base nitrogen rates are calculated for each of three classes. A crop reflectance sensor is then used during application to measure variability in crop condition within each class and to vary application up or down from the base rate • Maps of areas where broad-spectrum herbicide is to be applied are combined with reflectance sensors to deliver specific action chemicals to detected weeds	Variability across the whole field is sensed and used to control treatment rates during application. Examples: • Only green weeds, which are detected in fallow using reflectance sensors, are treated with herbicide • Crop reflectance sensors are used to control nitrogen application rate variation around a single base rate

Map-based management class operations can be undertaken with a wide range of application equipment and applied to any input that is metered onto a field. Real-time whole-field operations need specific hardware for the input being controlled. At present these types of operation are used only to apply nitrogen fertiliser, herbicide and growth regulants.

Deciding on the applicability and category of VRA operations for a situation can be done using a tree-structure of questions that require a positive or negative answer. An example is shown in Figure 7.7, where the decision begins with the premise that variability in crop yield is the initial signal that VRA might be warranted. Another model might begin with soil or landscape variability or variation in crop reflectance.

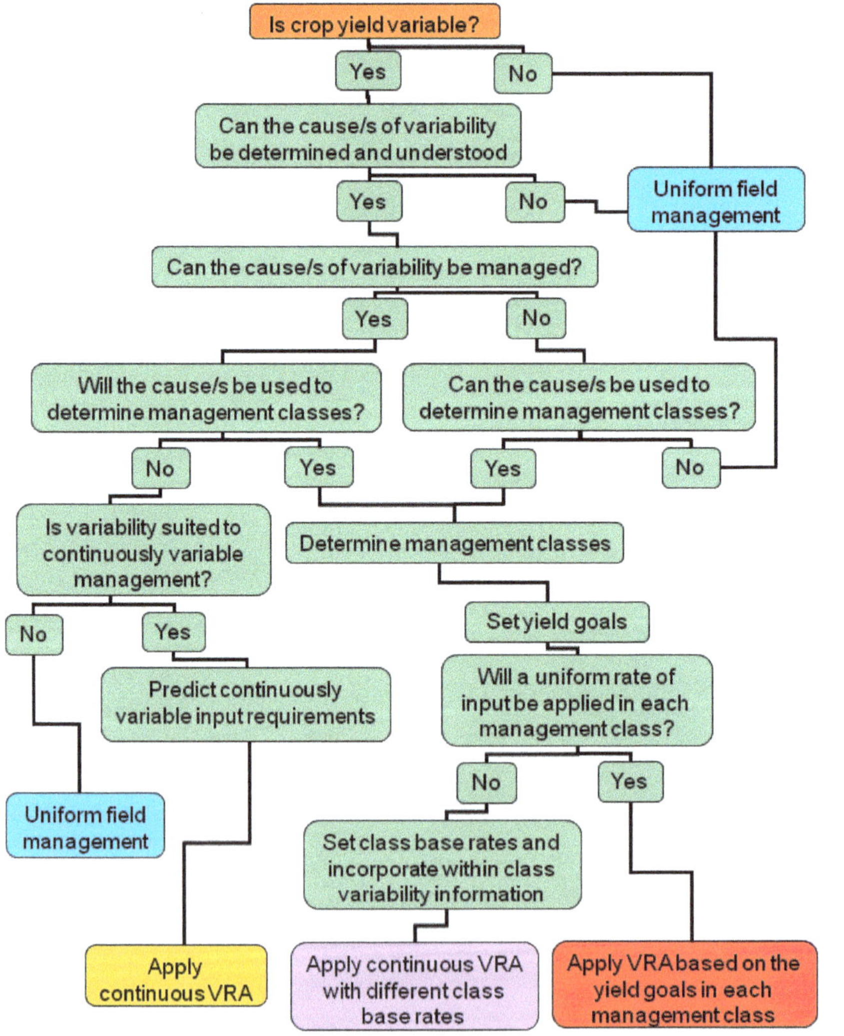

Figure 7.7: Basic decision tree for VRA management options based on a measure of crop yield variability.

SSCM decisions using maps of variability

Decisions direct from maps

Nutrient replacement

The amount of nutrient removed by a previous crop can be calculated using a yield map and a formula relating the amount of nutrient exported per tonne of grain. Using this process to prescribe fertiliser requirements for a following crop assumes that the optimum yield has been achieved across the field and/or that soil levels of the target nutrient were at or above critical levels prior to sowing the previous crop.

For example, the wheat yield map for a 110 ha field (Figure 7.8a) has been used to calculate the amount of phosphorus removed using the formula:

$$\text{P removed}_{(\text{kg P/ha})} = 4_{(\text{kg P/t})} \times \text{wheat yield}_{(\text{t/ha})} \qquad \text{(Eq. 7.1)}$$

The map produced (Figure 7.8b) can be used as the basis for phosphorus replacement rates in the field. However, it does not allow any margin for error in the estimate of how much phosphorus is removed per tonne of grain, nor the possibility that a base level of phosphorus may be required in the initial stages of crop growth. Equation 7.2 includes a base rate of 5 kg P/t to be applied over the whole field as part of the calculation of the total required phosphorus application rate (shown in Figure 7.8c).

$$\text{P removed}_{(\text{kg P/ha})} = 5_{(\text{kg P/t})} + 4_{(\text{kg P/t})} \times \text{wheat yield}_{(\text{t/ha})} \qquad \text{(Eq. 7.2)}$$

Both the maps in Figures 7.8b and 7.8c could be used for continuously variable application of phosphorus fertiliser. Alternatively, by dividing the field into high and a low management classes based on the phosphorus requirements, a two-rate application map can be made from the average phosphorus fertiliser requirements within the classes (Figure 7.8d).

For nitrogen in wheat, the simplest approach would be to take the yield data and use a uniform protein content as recorded by delivery dockets (or a number of samples if the crop is stored on-farm). The formula to be used is shown in Equation 7.3.

$$\text{N removed (kg N/ha)} = \text{yield (kg/ha)} \times \text{protein (\%)} \times 0.00175 \qquad \text{(Eq. 7.3)}$$

Figure 7.9b shows the mapped result for a 42 ha field of wheat (Figure 7.9a) using an average protein content of 14%. A more sophisticated and accurate approach would be to use spatial data gathered via an on-harvester protein sensor. Figure 7.10a shows a grain protein map as gathered by a Zeltex on-harvester protein sensor. The result of combining this data with the yield data using Equation 7.3 is shown in Figure 7.10b.

The nitrogen removed maps can be used to calculate continuous nitrogen prescription maps or the required nitrogen averaged for application with PMCs. Figure 7.11 is a two-PMC map calculated from the wheat yield, which shows the

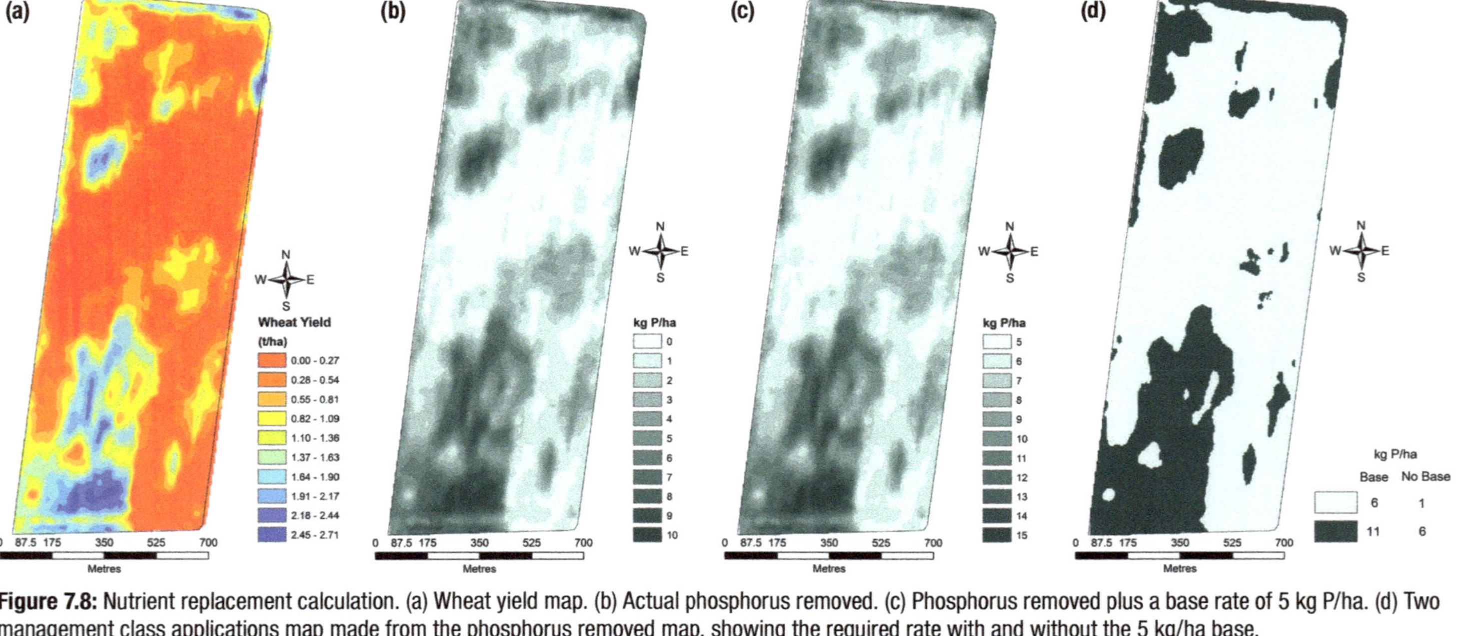

Figure 7.8: Nutrient replacement calculation. (a) Wheat yield map. (b) Actual phosphorus removed. (c) Phosphorus removed plus a base rate of 5 kg P/ha. (d) Two management class applications map made from the phosphorus removed map, showing the required rate with and without the 5 kg/ha base.

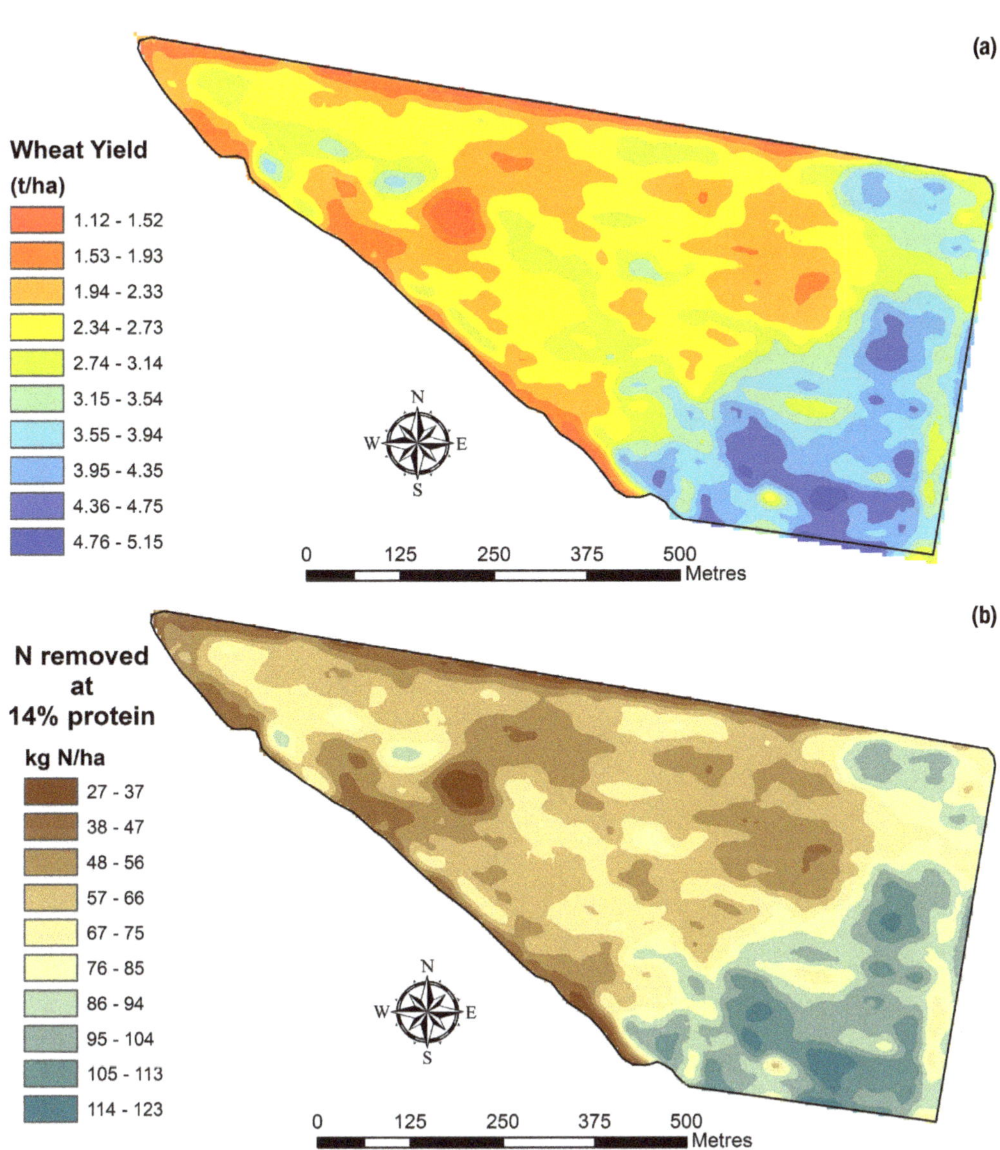

Figure 7.9: (a) Wheat yield map and (b) nitrogen removed (kg N/ha), calculated using a field average protein content of 14%.

average yield within each class. The values in Table 7.3 are a comparison of the nitrogen removed and the required replacement quantities of urea for each class, calculated from the maps in Figures 7.9b and 7.10b. In this instance it is clear that the estimate of 14% field average protein content that was used to calculate the nitrogen removed in Figure 7.9b was quite accurate. However, when using the true variation in protein as shown by Figure 7.10a, Class 1 is shown to require 3% more

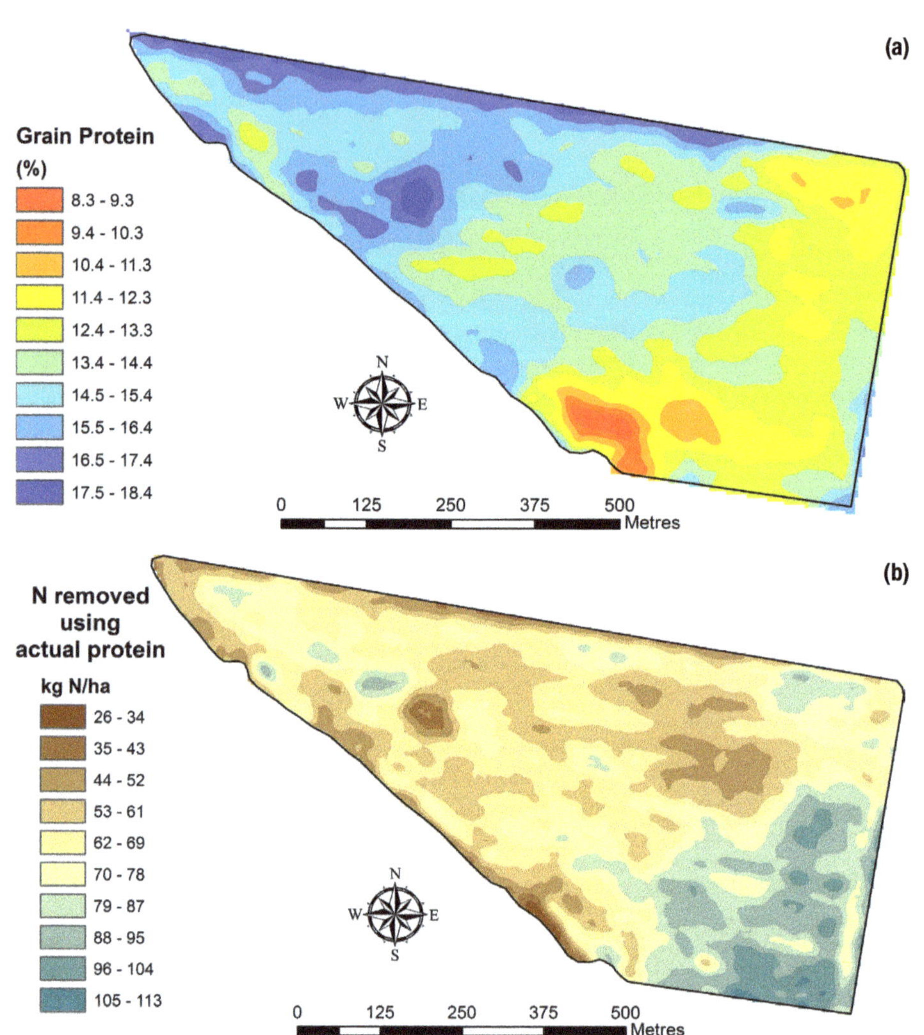

Figure 7.10: (a) Protein map gathered from an on-harvester monitoring system. (b) Nitrogen removed map calculated from the yield and protein data.

Table 7.3: Management class averages for output, nitrogen removed and replacement urea required under the two different protein information scenarios shown in Figures 7.9(b) and 7.10(b)

Class	Area (ha)	Average yield (t/ha)	Average protein (%)	Average nitrogen removed at 14% protein (kg N/ha)	Replacement urea required (kg/ha)	Average nitrogen removed at actual protein (kg N/ha)	Replacement urea required (kg/ha)
1	29	2.4	14.8	59	128	62	135
2	13	3.8	12.5	93	202	83	180
Total	42	2.9	14.1	70	151	68	149

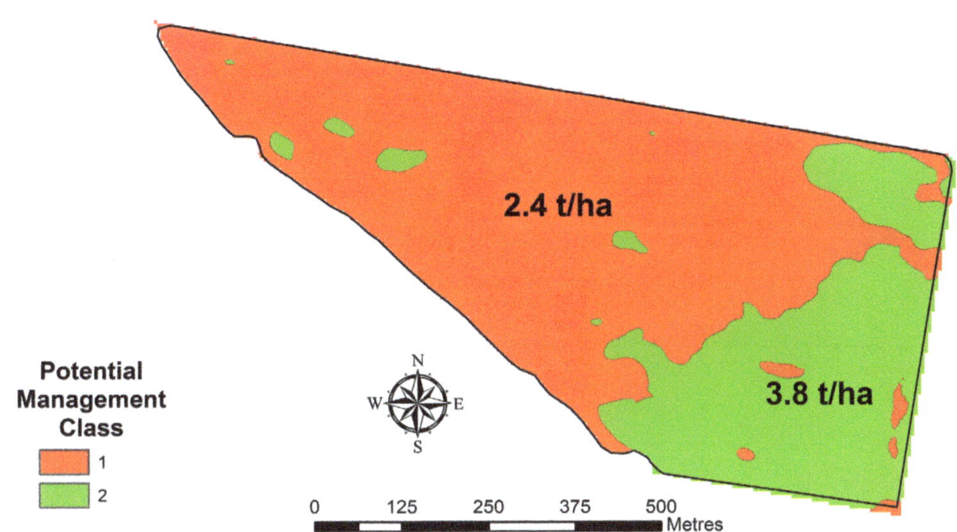

Figure 7.11: A two-management class map made from the crop yield data showing the average wheat yield for each class.

fertiliser than estimated by the simpler procedure, and Class 2 actually needs 12% less. While the more accurate procedure would result in only a small saving in total fertiliser used, the correct allocation to the classes would have agronomic benefits in future seasons.

Decisions from further investigation

When maps of variability are not directly linked to a management option, as in the previous examples, or the main causes of the variability are unknown or unquantified, then physical investigation of a field is necessary.

The patterns and quantities of yield variation documented in yield maps or calibrated crop imagery is usually the starting point for these further investigations for SSCM decisions. Mapping yield is also useful for assessing the results of changes in management and it enables the retention of data that may be useful in future decisions.

However, there are some circumstances where local knowledge of the interaction between soil/landscape resource variability and yield variability means that a map of soil variability or crop reflectance may be enough to identify sampling sites for investigation of specific local variability issues. These issues should be easily identifiable and have a significant impact on yield.

For example, in areas where variation in soil pH can typically dip low enough to limit crop growth, a soil EC_a map can be used to identify changes in soil characteristics. Sampling for soil pH at sites across the range of EC_a values allows the EC_a map to be calibrated for pH then used to drive VRA of lime ($CaCO_3$). In these cases, growers and their advisers need to be comfortable with their subjective assessment of the size of the initial yield variability, that the cause is correctly identified and that there is an ability to accurately assess the impact or value of management changes.

When there is no obvious agronomic problem causing the crop yield variability, further investigation using the maps of yield and resource variability is warranted. There are four general propositions to consider in these investigations:

- whether there is a transient, manipulatable factor that is restricting parts of the field from reaching seasonal yield potential;
- whether there is one (or a correlated combination of) static factor/s that dominates the changes in yield potential in a field;
- whether complex interrelationships between measurable factors need to be analysed to determine the cause of yield variation;
- whether the yield variability is caused by a change in the production process that was not measured (e.g. unobserved, localised pest damage or disease).

The first two scenarios simplify management decisions. Tackling the third requires the use of more complex modelling but may ultimately be optimal in terms of yield and environmental benefits. The fourth would probably show up in a correlation with a static factor unless there was a breakdown in normal standards of agronomic management.

The investigations would involve sampling or observations at locations to assess:

- soil physical and chemical conditions;
- crop condition and density;
- weed density;
- position in the landscape.

The aim is not to make a map of variability in these properties across the field, but to assess the size of any differences across the field and to use this information in decisions about changes to crop management.

Sampling strategies

Field sampling for this purpose has traditionally been based on gathering individual or composite samples from locations using a random sampling strategy. An improvement on this strategy is to use prior knowledge to guide the decision on the location of sampling sites. Maps that show variability across the whole field can be used singly or in combinations for this purpose.

Using a single map to direct sample site selection is possible, but is more risky than combining a number of map layers because it relies on one attribute and one set of seasonal conditions. Combining maps allows different components of the cropping cycle and/or different seasonal conditions to be included in the process of deciding where to sample.

A single map can be used to allocate samples across the full range of variability displayed, or used to create potential management classes within the field. When combining maps with different measurement units (e.g. t/ha, m, mS/m), dividing the field into PMCs is the best option.

Whether samples are to be allocated across a full range map or a PMC map, the number of samples to be taken must first be decided. This decision will ultimately be based on cost, but is also affected by:

- what is to be sampled;
- the procedure for obtaining the sample;
- analytical procedures.

Once the number of samples is determined, site allocation can take place. This will be based on:

- sampling evenly in quantiles across the distribution of values in a full range map. This attempts to describe the full range of variability in the field;
- random, replicated samples inside different PMCs with the constraint that boundaries are avoided. This attempts to estimate the difference between the classes and the variability within each class;
- sampling targeted at sites that possess close to the average values for all the attributes that were used to make the PMCs. If three different types of maps were used, a site would need to be close to average for all three properties for a particular PMC. This aims to better estimate the mean in each class using a small number of samples.

The depth of soil sampling and targets for crop sampling can be set to suit local agronomic testing regimes. Analysis of the sample data should provide some explanation of the causes of the observed yield variability and highlight any implications for crop management.

Using single maps

In the simplest case, a single map can be used to locate the sampling sites. The use of single-season yield maps is not widely recommended and should be attempted only if the map is known to be typical of the field and the investigation is targeting information for a specific management option (e.g. weeds, soil amelioration).

An example of a wheat yield map from a 300 ha field in NSW is shown in Figure 7.12. Soil samples were taken at points of low/medium/high values across the yield map as shown. The analysis of the samples provided valuable insight into the causes of the crop yield variability. The most obvious factor was the higher exchangeable sodium percentage (ESP) values (ESP >10, Table 7.4) in the soil at the lowest-yielding sites (3 and 6). Site 1 had a similar cation exchange capacity (CEC) and moisture content to those at the two highest-yielding sites (8 and 9) but a medium yield response. The elevated ESP at site 1 (ESP = 7.95) is depressing yield

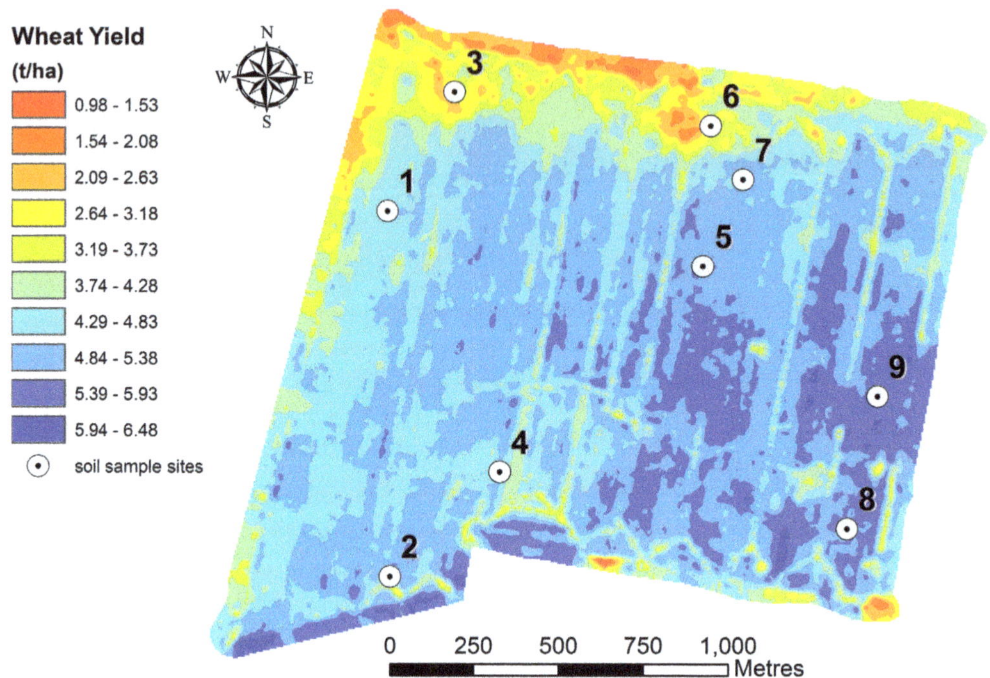

Figure 7.12: Wheat yield map and soil sample locations.

but not to the extent at sites 3 and 6. The medium-yielding area at site 4 was not restricted by ESP but it was in a depression that had the highest observed average profile moisture content, leading to the theory that some waterlogging had occurred.

If the soil sampling had been used to produce a single composite sample, then the average ESP value of 5.84% (Table 7.4) would have produced a requirement for a 1.3 t/ha gypsum application over the entire area (A$6358 @ $20 tonne gypsum). Sampling using the yield map pattern, however, makes it obvious that much of the field is not restricted by high ESP and thus requires no gypsum, and that the 1.27 t/ha would not be sufficient to ameliorate the poor areas.

The simplest way of dealing with this is to compartmentalise the poor area and apply gypsum at the average required from the enclosed sample points. This would result in the lowest-yielding 92 ha (average wheat yield = 3.5 t/ha) being treated at a gypsum rate of 5.8 t/ha (A$10 724) and 208 ha (average wheat yield = 5.0 t/ha) being untreated along the design shown in Figure 7.13.

Here the financial cost of the gypsum is higher but it is likely to benefit the crop yield into the future, whereas the whole-field application would provide little agronomic value and greater application costs. The three years following gypsum application are expected to show the largest impact of the treatment. Raising the

Table 7.4: Soil sample analysis across the yield pattern in Figure 7.12

Sample	Yield (t/ha)	Average profile moisture content (%)	CEC 0–30 cm	ESP 0–30 cm (%)	Bulk density 0–30 cm	Gypsum requirement (t/ha)
1	4.65	25.6	44.54	7.95	1.27	4.3
2	5.02	26.8	37.43	6.21	1.22	1.4
3	2.56	19.3	23.94	10.97	1.44	5.3
4	5.05	27.3	37.95	2.70	1.27	0
5	5.30	27.0	38.30	3.73	1.26	0
6	2.70	19.1	34.34	10.56	1.38	6.8
7	5.10	24.8	33.78	4.51	1.28	0
8	5.43	26.0	43.39	3.46	1.29	0
9	5.70	25.8	45.13	2.47	1.27	0
Field average	4.61	24.6	37.64	5.84	1.29	1.3

average yield in the wheat rotation year by 0.5 t/ha to 4 t/ha in the gypsum-treated 92 ha would increase returns by more than A$15 000 (wheat at A$260/t) for the initial A$10 724 gypsum outlay. Benefits in the two other seasons of the three-year crop rotation would add further financial value.

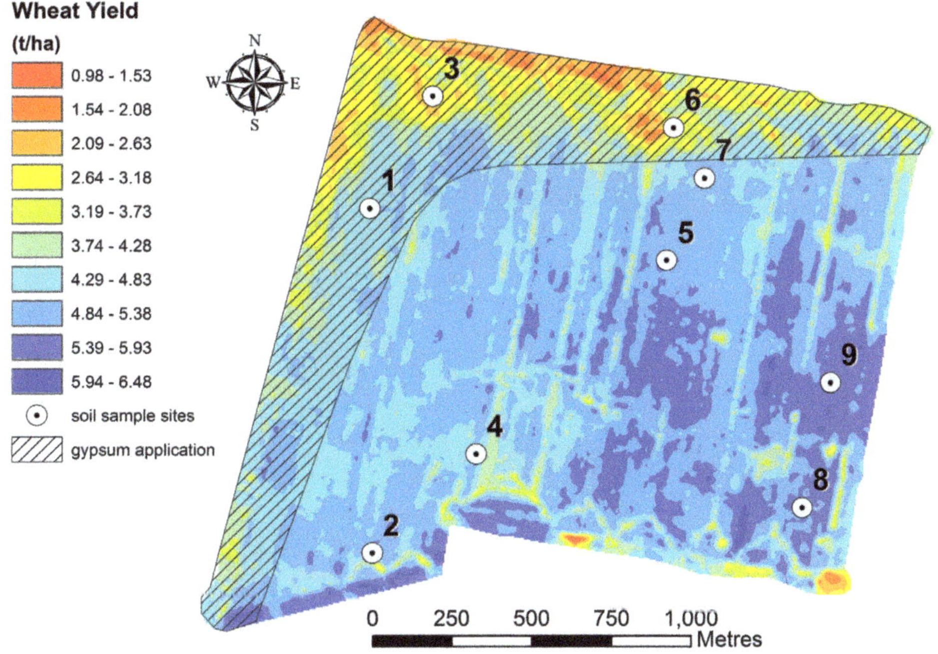

Figure 7.13: Possible gypsum application map based on the yield map sampling scheme.

Using PMCs

In this example, a 75 ha field has been divided into three PMCs using fine-scale EC_a, elevation and yield information. Soil sampling sites have been located within each of the three zones (Figure 7.14) in an attempt to explore causes of the yield differences between the classes. Soil cores were taken at each site and the samples divided into a topsoil (0–0.3 m) and subsoil (0.3–0.9 m) layer for analysis. The results are shown in Tables 7.5 and 7.6.

Analysis of the topsoil (Table 7.5) shows that Class 1 has a higher CEC and a lower sand fraction than Classes 2 and 3, but is lower-yielding (Figure 7.14). Of note for management is that the soil nitrate is also substantially higher in Class 1. An examination of the soil below 0.3 m (Table 7.6) shows that the CEC and clay content of Class 1 are significantly lower than in the other classes, and the soil nitrate remains double.

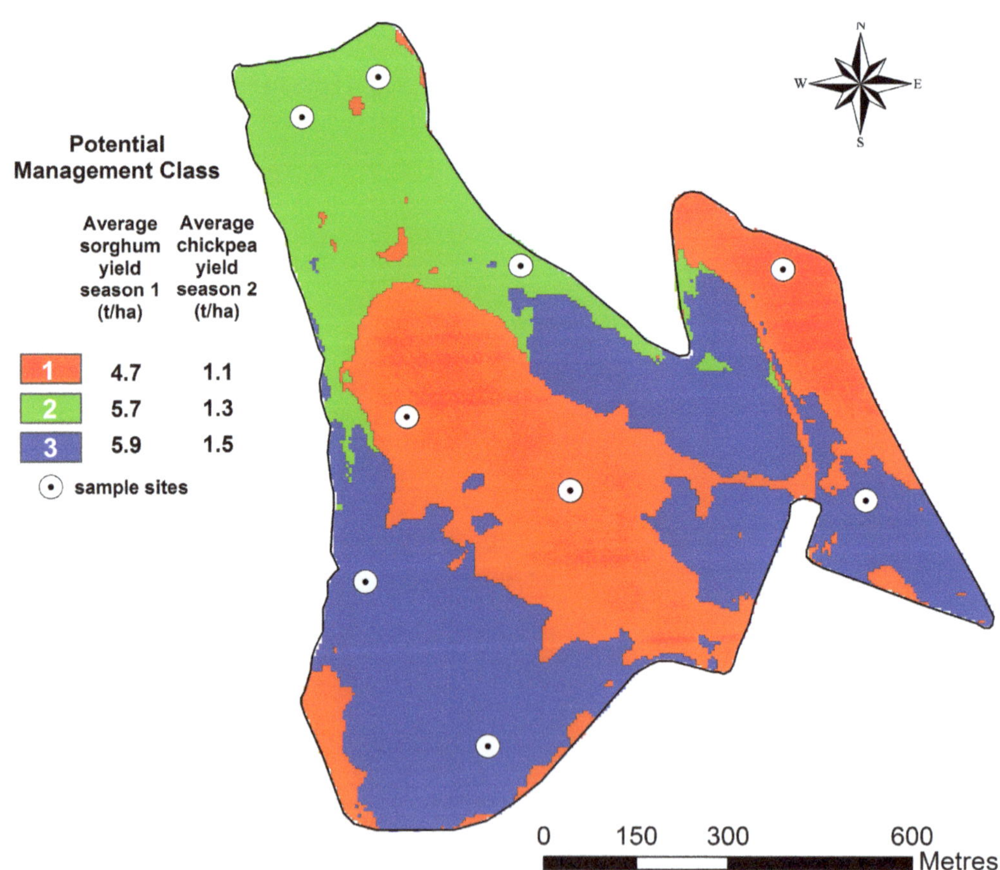

Figure 7.14: A 75 ha field divided into PMCs. Average yield responses for the PMCs for a sorghum and chickpea crop are shown in the legend.

Table 7.5: Soil test results for the 0–0.3 m soil layer at the sample sites in Figure 7.14

Soil attribute	Class 1 (red)	Class 2 (green)	Class 3 (blue)	Field mean
pH (CaCl)	7.6	7.2	7.8	7.5
O.C. (%C)	0.9	0.8	0.6	0.8
NO_{3^-} (mg/kg)	30.4	19.3	10.6	20.1
P (mg/kg)	5.3	6.3	2.7	4.8
K (meq/100 g)	0.6	0.9	0.5	0.7
Ca (meq/100 g)	62.6	40.5	51.3	51.5
Mg (meq/100 g)	13.2	18.3	22.1	17.9
Na (meq/100 g)	0.2	0.5	1.0	0.6
Total N (mg/kg)	1026	1079	658	921
CEC (meq/100 g)	77	60	75	70
Ca/Mg	4.8	2.2	2.3	3.0
ESP %	0.25	0.92	1.35	0.84
Sand %	10	16	12	13
Silt %	15	13	13	14
Clay %	75	71	75	74
E.C. (mS)	163	138	136	145

Table 7.6: Soil test results for the 0.3–0.9 m soil layer at the sample sites in Figure 7.14

Soil attribute	Class 1 (red)	Class 2 (green)	Class 3 (blue)	Field mean
pH (CaCl)	7.7	7.8	8.0	7.8
O.C. (%C)	0.8	0.7	0.6	0.7
NO_3^- (mg/kg)	14.7	11.9	5.6	10.7
P (mg/kg)	3.7	3.0	2.5	3.1
K (meq/100 g)	0.42	0.65	0.48	0.5
Ca (meq/100 g)	42.1	38.9	47.0	42.7
Mg (meq/100 g)	9.5	21.5	24.9	18.6
Na (meq/100 g)	0.3	2.1	2.7	1.7
Total N (mg/kg)	887	687	532	702
CEC (meq/100 g)	52.3	63.4	74.8	63.5
Ca/Mg	5.2	1.8	1.9	3.0
ESP %	0.7	3.2	3.6	2.5
Sand %	18	15	11	15
Silt %	17	11	11	13
Clay %	65	74	78	72
E.C. (mS)	126	162	155	148
Soil depth (m)	0.68	1.17	1.24	1.03
Profile avail. H_2O at sampling (mm)	68	108	128	101

The interpretation of these results is that the difference in the physical properties of the subsoil, combined with the fact that the soil is on average 40% shallower in Class 1 (Table 7.6), conspires to restrict the quantity of available moisture in the profile of Class 1 compared to Classes 2 and 3. This relative limitation in soil moisture in Class 1 would limit crop yield during most seasons and therefore reduce the nitrogen requirement. Under uniform fertiliser management, accumulation of soil nitrogen reserves in the whole profile (as evident in nitrate and total N levels in Tables 7.5 and 7.6) would then occur.

While there is also some evidence for the build-up of phosphorus in the topsoil of Class 1, from a management perspective it is the manipulation of soil nitrogen that would bring the biggest benefit to the profitability of the field.

Varying input rates between management classes

Having found the major drivers for the spatial variability between PMC in a field, the possible changes to input management can be explored. Three general options for determining input rate changes are based upon:

- maintenance of a nutrient balance to achieve a uniform yield goal;
- modifying yield goals between classes;
- rate response experiments across classes.

The choice of input in each field will be made on the basis of the results obtained from strategic sampling within the PMCs. A marked build-up or depletion in a soil nutrient between classes could be used as a criterion, along with the magnitude of contribution the required input makes to the variable costs of production. Managerially significant differences in soil physical attributes or restrictions or crop growth/disease parameters that are identified during the investigation operations may also drive changes in relevant inputs across a field.

Maintenance of a nutrient balance to achieve a uniform yield goal

This option involves the establishment of a yield goal for a field, calculating the nutrient requirements for the goal, estimating the quantity of nutrient already available in the soil and determining the additional amount required to achieve the goal. Using this traditional agronomic technique for calculating nutrition requirements provides a simple method for including the information obtained from investigative sampling within PMC into input rate decisions.

Equation 7.4 is often used to calculate the nitrogen requirements for wheat crops:

$$\text{N required (kg N/ha)} = \text{yield goal} \times \text{protein goal} \times 1.75 \times 2 \qquad \text{(Eq. 7.4)}$$

According to Equation 7.4, to achieve a yield goal of 4.5 t/ha at 13% protein would require 205 kg N/ha to be available to a wheat crop during the growing season. As an example, consider the field in Figure 7.14. Applying the 4.5 t/ha yield goal uniformly across the field, and using the soil nitrogen information in

Table 7.7: Available nitrogen in the soil and the calculated nitrogen requirements for three PMCs. Comparison between uniform application based on the average available soil nitrogen content and variable-rate treatment based on actual available soil nitrogen in each of the PMCs

	Class 1 (29 ha)	Class 2 (13 ha)	Class 3 (33 ha)	Field average (75 ha)
Average available N in the soil (kg N/ha)	238	172	86	165
Uniform N application (kg N/ha)	40	40	40	40
Actual N required to obtain 205 kg N/ha in soil (kg N/ha)	0	33	118	
Consequence of uniform N application	2.5 t urea waste	197 kg urea waste	66% underfertilised	

Tables 7.5 and 7.6 as pre-sowing sampling information, the nutrient balance requirements can be calculated.

First, the nitrogen values from the 0–30 cm and 30–60 cm soil samples are used to calculate the average amount of nitrogen in kg/ha that is resident in the soil profile. Table 7.7 shows the results of these calculations for each class and the overall average for the field. The overall average is the estimate that would be traditionally used to calculate the uniform fertiliser requirement for the field. In this instance, the field average of 165 kg N/ha would suggest that a further 40 kg N/ha is required across the field to complete the yield goal nitrogen balance (Table 7.7). In this conservative approach, no additional contribution of nitrogen from mineralisation in the soil over the growing season is included.

However, because the soil sampling has been done in classes, the class average resident nitrogen can be used to calculate the additional nitrogen required to complete the yield goal nitrogen balance in each class. The results (Table 7.7) suggest that very different amounts of fertiliser would be required within each class.

Applying the uniform treatment rate would overfertilise Classes 1 and 2 and underfertilise Class 3. A total of 2.7 t of urea would be wasted by the uniform application and the substantial underfertilisation in Class 3 would result in yield and protein losses in an average or better season. Varying the fertiliser rates to achieve the 205 kg N/ha for the 4.5 t/h yield goal in each class would be a more profitable and less environmentally risky management decision.

Modifying yield goals between classes

Sampling, then modifying fertiliser rates to reach an even mass balance across a field, does not acknowledge that in many cases the yield potential is different between the management classes. Following directed sampling to identify the local factors affecting yield variability, new yield goals can be assigned to the different PMCs. This can be achieved by combining the information on the differences in the soil/landscape properties between classes with local agronomic advice on yield

Table 7.8: Yield differences between PMCs in Figure 7.15

Class	Average wheat yield (t/ha)		
	Season 1	Season 2	Combined
1	2.4	2.6	2.5
2	3.8	3.2	3.5
Whole field	2.9	2.8	2.9
Difference between Classes 1 and 2 (%)	37	19	28

potential. Alternatively, the relative yield potentials can be assessed using the yield differences shown in the previous crop yield maps.

Either way, local experience should be included to set a maximum yield goal for the highest-performing class. The yield goals for the other classes should decrease on a percentage basis.

Figure 7.15 shows three data layers (a = EC_a, b = elevation and c = wheat yield) from a 40 ha field that were used to derive two PMCs (d). Table 7.8 contains average yield data figures for the whole field and each PMC. The wheat yield of Class 1 is 19–37% (average 28%) less than Class 2. Once management is convinced that there are no remaining amelioration questions and that the yield differences are repeating over seasons, they could consider reducing the yield goal and the input rates in Class 1 accordingly. There are a number of available options:

- set a yield goal for Class 2 based on local advice and reduce the yield goal of Class 1 by either the minimum/average/maximum percentage difference seen over the seasons. The most risk-averse option is to reduce the yield goal by the minimum percentage. Where three classes are being used, a further option is to set the yield goal for the middle class based on local calculations, and vary the goals up and down respectively for the other classes;
- set yield goals based on the actual maximum yields obtained in each class;
- set yield goals based on the actual averages obtained in each class.

This general procedure can be adapted for more classes and modified for different inputs and crops.

Rate response experiments within classes

Setting up some basic experiments across a field can provide a more comprehensive understanding of the actual response to different application rates. Where PMCs have been identified, field-scale experiments can be established to estimate the response in each PMC to a single input.

It is best to keep these types of experiments to testing changes in a single input or practice. It is preferable to increase the number of different rates in the experiment, rather than include additional variables at this scale.

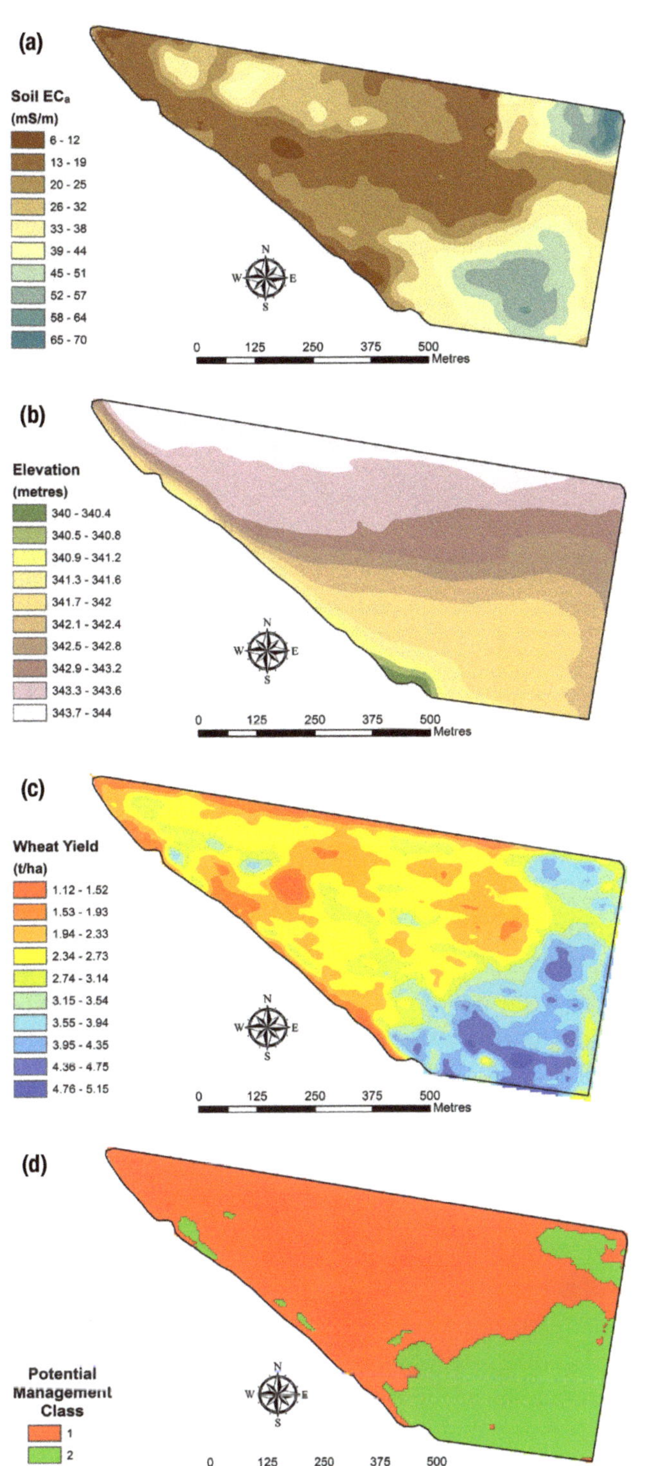

Figure 7.15: (a) Soil EC_a, (b) elevation, (c) crop yield and (d) two-class PMC map for a 40 ha field.

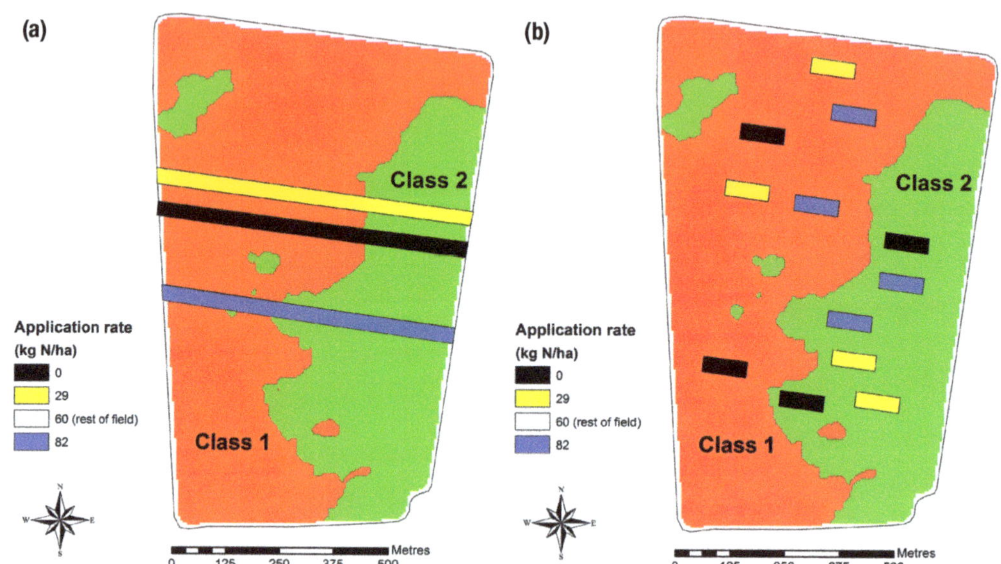

Figure 7.16: Two general options for rate response experiments in a field with two PMCs. (a) Whole-field strips. (b) Small strips.

The design of the experiments should consider application equipment capability and width, incorporation of the management class pattern and the desirability of minimising the area and financial impact of the experiment. These experiments can be run using whole-field treatment strips (Figure 7.16a), but where variable-rate technology is available, the application of small strip experiments is easy (Figure 7.16b). The use of small strips allows correct independent replication, more flexibility in location and greater coverage of the conditions in the field, and often results in less area being subjected to potential yield-reducing treatments.

Replication helps to ensure the effect of a particular treatment is not affected by other sources of variation. For example, in an experiment where a field is divided in half and a treatment is applied to one half and a control to the other, it is impossible to tell if any difference in response reflects real treatment effects rather than inherent differences between the two halves. Replicating treatments across the field would help solve this problem. Randomisation of treatment rates to treatment strips or plots ensures that any treatment effects are representative of the experimental area and not biased by choosing to put high or low rates in a particular spot.

The majority of the field should be treated with the control application (i.e. traditional best practice), while alternative treatments could include multiples of the traditional application rate (e.g. 0.25, 0.5, 1.5, 2). More than one alternative treatment rate is preferable, otherwise the best that can be hoped for is a decision that one treatment is preferable to another. At least two alternative treatments can help provide information to decide the 'best' treatment rate.

The information from an experiment will be greatest when there is a good size difference between the treatment rates. Where possible, a zero rate treatment should be included to help show the full scale of response. This is especially important when the experiment is testing whether a reduction in input rate is viable. Obviously, placing zero rates in the field can cause some trepidation, but keeping the size of the treatments small can help ease such concern. With these designs, data from the whole field can be used in the analysis.

Whichever method is chosen, the main requirements are:

- treatments must be laid out in the direction of sowing and harvesting;
- each treatment application should be at least three harvester widths wide to ensure that at least one full cutting width can be obtained from each strip without the possibility of contamination from adjoining treatments. Therefore, the minimum strip width will be controlled by the minimum multiple of the application machinery width that will meet this target;
- the minimum length of each treatment plot will be constrained by the operational mechanics of the harvesters. With grain mixing within the harvester occurring along the direction of operation, yield data gathered at the beginning and end of each strip should be regarded as contaminated by surrounding treatments (usually standard field or headland treatment). The plots should be a minimum of 100 m long, which would ensure most mechanical set-ups are covered. It is suggested that data from the first and last 20 m of each treatment plot be discarded from the eventual analysis.

Analysis

Numerous procedures for analysis can be undertaken, with varying degrees of statistical legitimacy. The 'standard' statistical analyses are used to test whether the average results of different treatments are equal. The two commonly used tests are:

- t-test – used to compare two different treatments;
- analysis of variance (ANOVA) – applied to test for differences between more than two treatments. When used to compare two treatments, it is the same as the t-test.

Both these tests assess the overall variation in the response data and calculate how much can be attributed to effects from the treatments and how much is occurring between the individual measurements for each treatment. The result is expressed as whether or not there is a 'significant difference' between the treatment averages.

The finding of a 'significant difference' is an assessment that the average results from the two treatments are statistically different enough that the probability of such a difference happening by chance is below a set threshold. The thresholds are known as 'levels of significance'. Popular levels of significance are 5% ($p = 0.05$),

1% (p = 0.01) and 0.1% (p = 0.001). For example, if someone argues that 'there is only one chance in a thousand this could have happened by chance', a 0.001 level of statistical significance is being implied. The lower the significance level, the stronger the required evidence for a difference. Choosing a level of significance is an arbitrary task, but for many agricultural applications a level of 5% is chosen.

There are a number of issues with using these type of tests in SSCM. The tests use average treatment response for comparisons, and include an assessment of the accuracy of the averages when making a decision on significance. The accuracy is assessed by the amount of variation in all the measurements and by the number of measurements used to make the averages. In these tests, as the number of measurements increases, the accuracy of the estimates for the averages and the overall variation are deemed to increase and the easier it is to find significant differences.

However, each measurement is expected to be from an independent sample of the property being measured. One of the main points that is overlooked in applying common statistical analysis in SSCM is that the individual data points collected by a yield monitor are not statistically independent sources of information. The way the harvester applies internal mixing of grain means that neighbouring samples taken by the monitor are correlated, so each measurement does not constitute a new piece of information in a statistical sense. Given that yield monitors can collect a lot of measurements in most treatment plots, if this is not recognised and each measurement is considered independent in a statistical analysis, then the chance of finding statistically significant treatment effects is artificially inflated simply by using a yield monitor to harvest the experiment.

One way to improve the reliability of the results obtained from using these tests on strip experiments is to break the response data into sections and analyse the average response values in each section. When the response measurements are yield monitor data, the sections can be created by imposing 20 m buffers along the strip at 50 m intervals (Figure 7.17). While this does not absolutely address the statistical issues, using averages of separated sections as the measurements allows the number of observations to be reduced, which helps make the accuracy assessment more realistic.

It is also worth considering that the discovery of a statistically significant difference does not necessarily equate to an agronomically significant difference. The reverse is also true. The fact that a statistically significant difference is not found, does not necessarily mean that an agronomically significant difference is not present.

The results of using yield monitor data in standard statistical tests can provide information for decisions on rate changes, but they should be considered with caution.

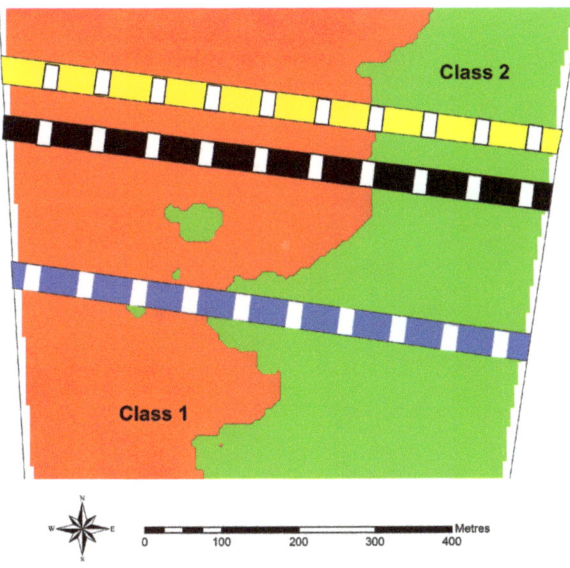

Figure 7.17: Whole-field treatment strips showing imposed segmentation.

For multi-rate experiments (three or more treatments), it is possible to provide an agronomic assessment of the optimum rate from the experiment. In this case it is not the maximum yield that is sought, but the yield that optimises the cost of production and financial returns. For this analysis the cost of the input and the price received for the crop yield is included in a marginal rate analysis, which compares the marginal revenue (MR) to the marginal cost (MC) of a production scenario. MR and MC are defined as either the change in revenue or cost as each additional unit is produced. Here, MC is the price of each unit of the input being applied (e.g. the price of 1 kg of nitrogen) and MR is the amount of money received for the output (crop yield) gained from using that unit of input.

When the MR for applying an additional unit of input is greater than the MC, applying the unit of input is profitable. If MR is less than MC, the additional unit of input is not profitable. Since total profit increases when marginal profit is positive and total profit decreases when marginal profit becomes negative, it must reach a maximum where marginal profit is zero – or where MC=MR. This is because the producer has collected positive profit until the point where MC=MR.

A nitrogen fertiliser case study

Figure 7.18 shows the actual fertiliser application map and the subsequent yield for the nitrogen response trial designed in Figure 7.16b. The average rate for the field was 60 kg N/ha. The average yield from each treatment plot was calculated, along with the average for each treatment rate in each class. The values are shown in Table 7.9.

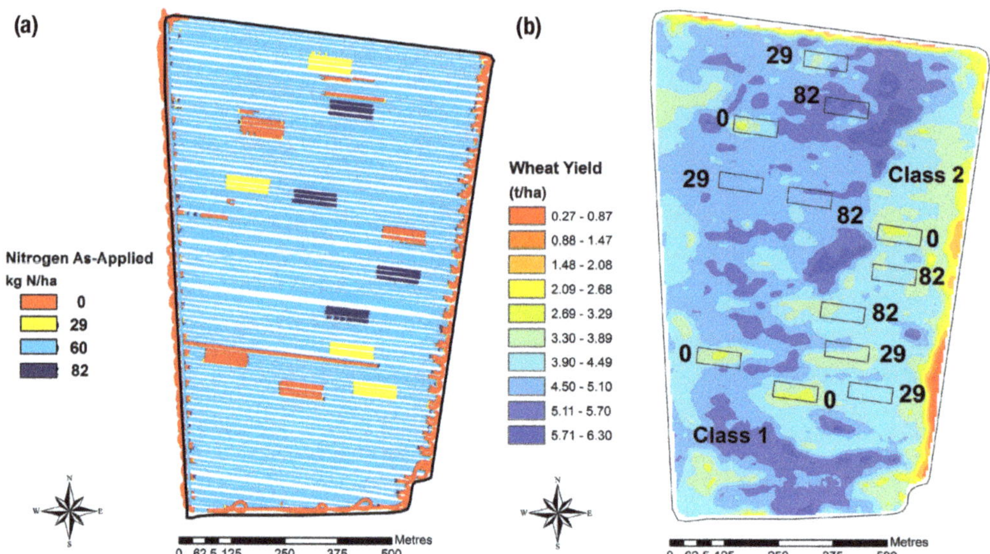

Figure 7.18: (a) The as-applied map for the nitrogen response experiment shown in Figure 7.16b where the uniform field treatment was 60 kg N/ha. (b) The subsequent wheat yield map.

The results show that the two classes are responding differently to the applied nitrogen fertiliser and that the uniform rate may not be best for either class. The season was above average in rainfall, with growing-season rain (June–November) of 350 mm (in the 70th centile).

Figure 7.19 shows the results of ANOVA analysis using individual yield values obtained from the harvester in each trial strip. In Figure 7.19a all the treatments are considered statistically different when the trial is analysed as a whole. When the field is analysed in management classes, all the treatments are considered statistically different in Class 1. Class 2 shows that 0 kg N/ha and 60 kg N/ha are statistically different from each other and from the 29 kg N/ha and 82 kg N/ha treatments. The response to the 29 kg N/ha and 82 kg N/ha treatments are not statistically different from each other.

When the average yield values for the trial strips are used in the same tests, the results are less decisive (Figure 7.20). In the whole-field and management class

Table 7.9: Average yield values for the treatment strips in Figure 7.18

Rate (kg N/ha)	Class 1			Class 2		
	Rep 1	Rep 2	Average	Rep 1	Rep 2	Average
0	3.32	3.76	3.54	3.41	3.25	3.33
29	4.68	4.15	4.41	4.09	4.07	4.08
60	4.77	4.90	4.84	4.00	3.95	3.97
82	4.97	5.50	5.24	4.11	4.16	4.13

analyses, only the response to the 0 kg N/ha treatment can be considered statistically different from the others. The change in the statistical difference assessment is due to the decrease in the number of measurements used, the degrees of freedom (DF) in the analysis and the resultant effect on the calculated accuracy of the averages in the test procedures.

These results highlight the potential issues with using the 'standard' statistical tests with this type of practical experimental data. Figure 7.19, where each yield data point is treated as an individual measurement, finds statistical differences between yield values that would probably not be considered agronomically different (4.486 t/ha and 4.684 t/ha, Figure 7.19a). When the data is reduced to averages for each treatment and the perceived number of replications is reduced, the reverse is the case. Differences in yield response that would be considered agronomically significant are not identified as statistically different; for example, the 0.82 t/ha difference in yield between the 29 kg N/ha and 82 kg N/ha treatments in Class 1 (Figure 7.20b).

These issues aside, it is possible to identify which of the four rates produces the most yield from the trials using ANOVA, but not the rate that optimises yield and nitrogen input. To identify the optimum application rate from the trial, the yield response functions for the two classes using the average yield data from the strips were compiled (Figure 7.21a). The MR = MC analysis (Figure 7.21b) using prices for the season shows that the optimum rates of applied nitrogen fertiliser are 109 kg N/ha and 39 kg N/ha for Classes 1 and 2 respectively.

With this information, it is possible to compare the gross margin (GM) under standard management to the GM achievable under optimum-rate management. The GM is calculated by multiplying the yield by price and deducting the cost for the quantity of fertiliser applied. Where the GM of the standard management is less than the GM of optimum-rate management, the difference is termed a 'net wastage' of standard management. In the opposite scenario, the difference would be a 'net gain' for standard management.

Two simple management response scenarios can be considered, to make use of this information on variability. Scenario 1 would maintain the total amount of fertiliser applied to the field but moves the overapplication on Class 2 to Class 1. This would keep fertiliser costs constant and result in yield gains in Class 1 that would improve the gross margin by A$11.50/ha across the field. Scenario 2 would aim to apply the optimum amount to each class, requiring an additional 1.4 t of nitrogen to achieve the yield goals of 5.4 t/ha in Class 1 and 4.0 t/ha in Class 2. However, the increased yield would mean that the gross margin for the field would be improved by A$25/ha.

These figures are a function of fertiliser usage and prices, along with crop sales prices, in the specific year of each trial. To provide a simple way of standardising the wastage figures between other fields and subsequent years, the total net wastage can be compared to the specific nutrient fertiliser bill for the individual field

(a)

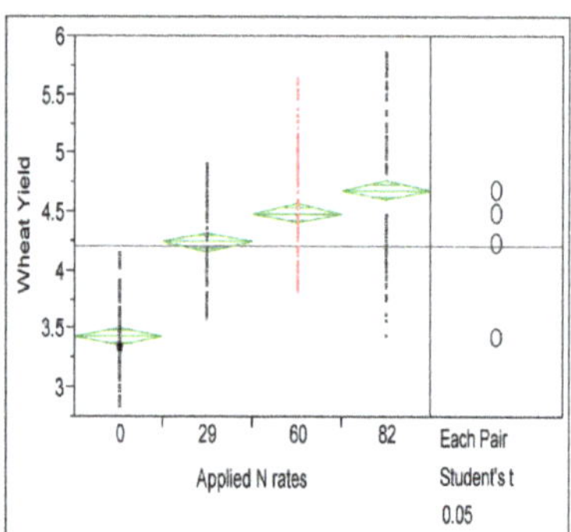

Analysis of Variance (ANOVA) – Whole field

Source	DF	Sum of Squares	Mean Square	F Ratio	Prob > F
applied N rate (kg/ha)	3	119.13	39.71	181.72	<.0001*
Error	524	114.51	0.22		
Total	527	233.64			

Comparison of means (student's t-test)

Rate					Mean (t/ha)
82	A				4.684
60		B			4.486
29			C		4.246
0				D	3.435

Rates not connected by same letter are significantly different.

(b)

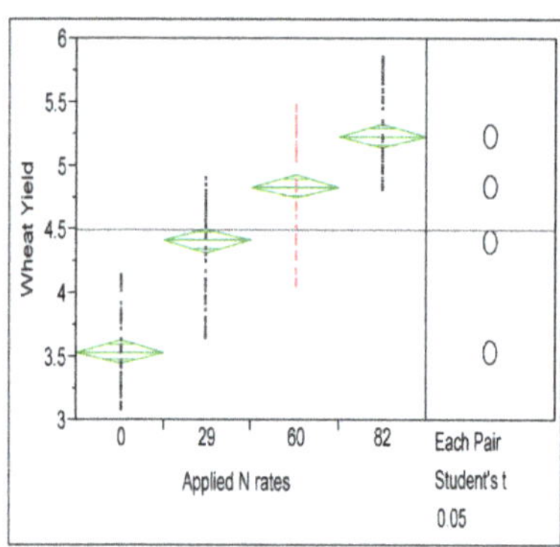

Analysis of Variance (ANOVA) – Class 1

Source	DF	Sum of Squares	Mean Square	F Ratio	Prob > F
applied N rate (kg/ha)	3	96.40	32.13	229.58	<.0001*
Error	261	36.53	0.14		
Total	264	132.93			

Comparison of means (student's t-test)

Rate					Mean (t/ha)
82	A				5.236
60		B			4.835
29			C		4.415
0				D	3.541

Rates not connected by same letter are significantly different.

(c)

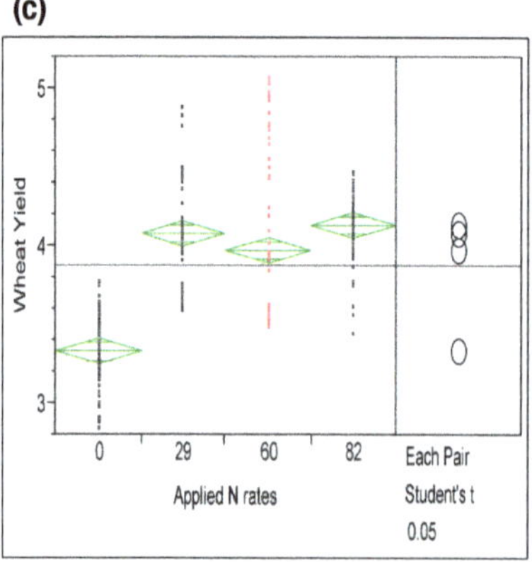

Analysis of Variance (ANOVA) – Class 2

Source	DF	Sum of Squares	Mean Square	F Ratio	Prob > F
applied N rate (kg/ha)	3	41.56	13.85	113.24	<.0001*
Error	259	31.69	0.12		
Total	262	73.25			

Comparison of means (student's t-test)

Rate				Mean (t/ha)
82	A			4.132
29	A	B		4.079
60		B		3.975
0			C	3.336

Rates not connected by same letter are significantly different.

Figure 7.19: One-way ANOVA of wheat yield (t/ha) by applied nitrogen rate (kg N/ha). Individual yield data from the trial plots used as input in the analysis of the trial response over (a) the whole field, (b) management Class 1 and (c) management Class 2.

(a)

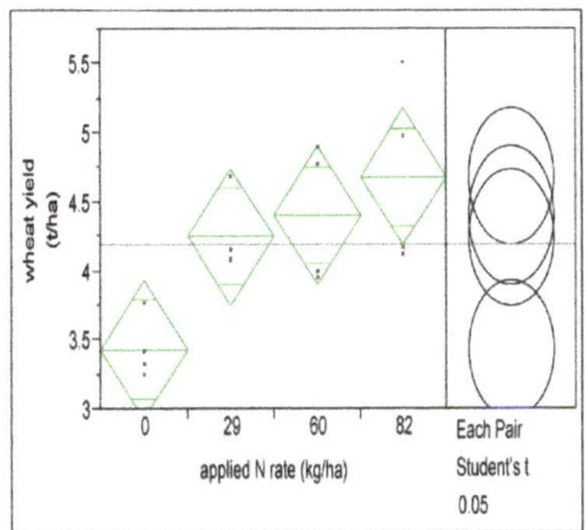

Analysis of Variance (ANOVA) – whole field

Source	DF	Sum of Squares	Mean Square	F Ratio	Prob > F
applied N rate (kg/ha)	3	3.46	1.15	5.52	0.0129*
Error	12	2.51	0.21		
Total	15	5.97			

Comparison of means (student's t-test)

Rate			Mean
82	A		4.685
60	A		4.405
29	A		4.247
0		B	3.435

Rates not connected by same letter are significantly different.

(b)

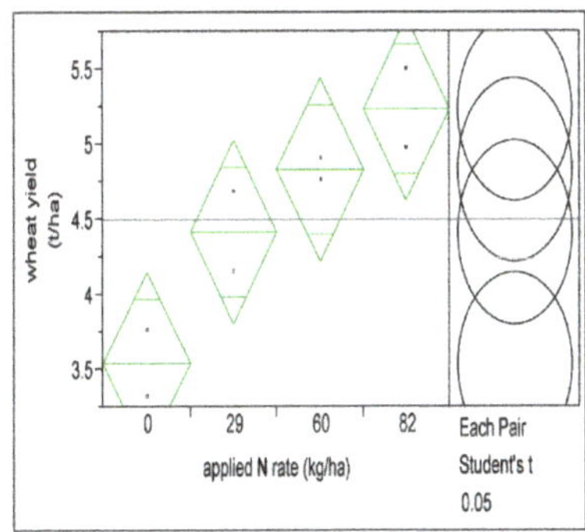

Analysis of Variance (ANOVA) – Class1

Source	DF	Sum of Squares	Mean Square	F Ratio	Prob > F
applied N rate (kg/ha)	3	3.16	1.05	10.92	0.0214*
Error	4	0.39	0.096		
Total	7	3.55			

Comparison of means (student's t-test)

Rate			Mean
82	A		5.235
60	A		4.835
29	A		4.415
0		B	3.540

Rates not connected by same letter are significantly different.

(c)

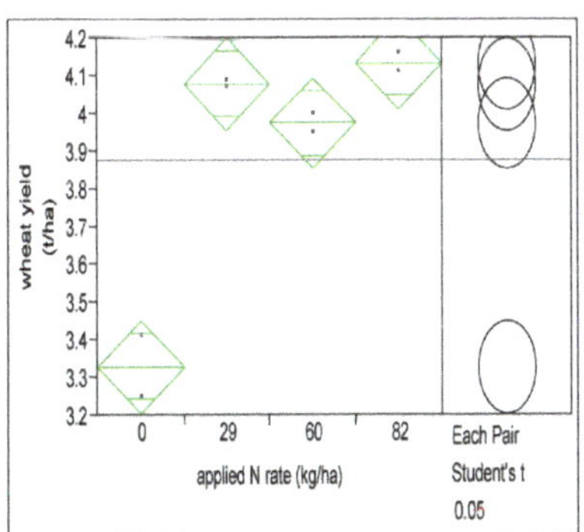

Analysis of Variance (ANOVA) – Class2

Source	DF	Sum of Squares	Mean Square	F Ratio	Prob > F
applied N rate (kg/ha)	3	0.83	0.28	71.665	0.0006*
Error	4	0.016	0.0039		
Total	7	0.85			

Comparison of means (student's t-test)

Rate			Mean
82	A		4.135
60	A		3.975
29	A		4.080
0		B	3.330

Rates not connected by same letter are significantly different.

Figure 7.20: One-way ANOVA of wheat yield (t/ha) by applied nitrogen rate (kg N/ha). Average yield data from the trial plots used as input in the analysis of the trial response over (a) the whole field, (b) management Class 1 and (c) management Class 2.

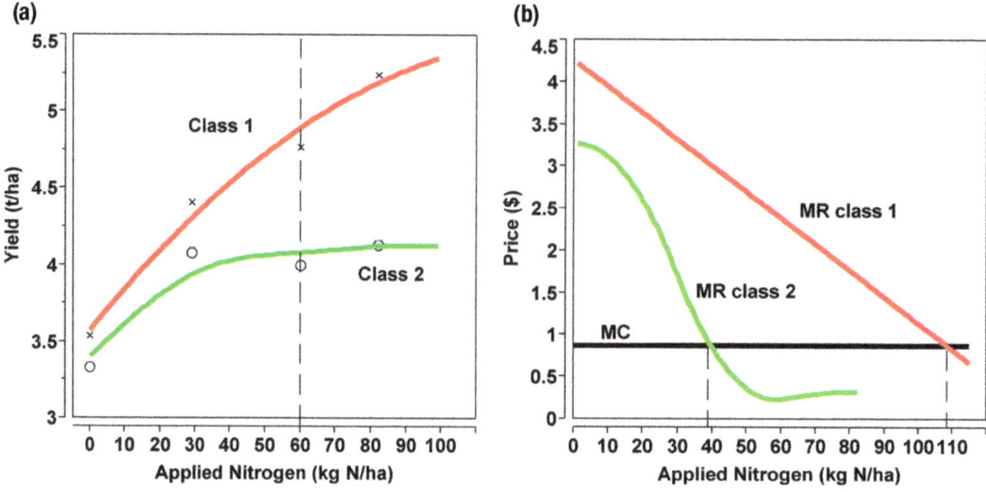

Figure 7.21: (a) Yield response to applied nitrogen fertiliser from the trial in Figure 7.18a where the uniform field application was 60 kg N/ha. (b) An MC = MR analysis shows that Class 1 optimum = 109 kg N/ha and Class 2 optimum = 39 kg N/ha.

average application in that year. In this case the calculation would estimate the wastage as a proportion of nitrogen fertiliser input costs, i.e. the investment in nitrogen fertiliser each year. It is calculated as a ratio:

$$\text{Proportion of fertiliser costs (\%)} = \frac{\text{nett wastage (\$)}}{\text{fertiliser bill (\$)}} \times 100 \qquad \text{(Eq. 7.5)}$$

Applying Equation 7.5 to this experiment shows that the proportion of nitrogen fertiliser costs potentially wasted by using standard management are 22% and 45% for scenarios 1 and 2 respectively. For more examples, see Chapter 8.

More-sophisticated approaches to the analysis of these types of spatial experiments require more advanced statistical or model-based approaches. These may be too difficult at present to apply on-farm, but are being developed for use in future software.

SSCM decisions using real-time sensors

Ultimately, the assessment and treatment of variability would be undertaken in real time and the scale of treatment effectively restricted only by the functional abilities of the application equipment (i.e. continuously variable real-time application).

As of 2013, the only widely available commercial applications of this concept in Australia are the use of real-time crop reflectance/nitrogen application instruments and real-time weed treatment technologies. They all rely on calibrating a treatment procedure to the measured reflectance of light from plants.

The simplest use is in treating weed plants in a fallow field. The process identifies green plants on a brown (soil or crop stubble) background and applies a target herbicide only in areas where living plants are identified. The situation is more complicated if there are weeds in growing crops; different plant types or off-row locations need to be identified before treatment can be activated.

The use of reflectance measurements for nitrogen management relies on calibrations between reflected light and crop biomass and/or nitrogen content. In general, the management response is based on detecting how well the crop is growing and calculating a suitable nitrogen application rate. The critical points to be considered with this approach are:

- whether the calibrations are accurate;
- whether the variability in growth is mainly due to variability in nitrogen availability or to some other factor;
- whether varying nitrogen application across the field will lead to a variable response from the crop.

The general operational characteristics of the main commercial systems for variable-rate nutrient and herbicide application are as follows. Readers are advised to consult the manufacturers or local distributors to check for updates to the operation of these systems.

Crop Circle™

Aim

To determine changing nitrogen application requirements in-season and provide VRA prescriptions.

Technique

The system incorporates three optical measurement channels that can be adjusted in-field to select optical measurement bands of interest, thereby allowing the instrument to be spectrally customised to a particular sensing application. Spectral configuration is performed via the use of standard 12.5 mm interference filters.

The sensor is calibrated to determine an index related to the mass of green crop (green area index (GAI)). This index is then compared to the expected index based on the crop variety and stage of development. Crops with a low GAI for the development stage are identified as requiring more nitrogen and those with a higher than expected GAI receive reduced nitrogen. Areas with the required GAI receive the planned nitrogen application. The planned nitrogen application is set by the user.

It is also possible to operate the system using a 'virtual reference strip'. The sensors measure the average GAI across a transect of the field and the planned nitrogen application is set to that value. Nitrogen is then varied up or down across the field depending on changes in the observed GAI. Again, the user has total control over the planned amount of nitrogen applied, whether more nitrogen is applied with lower GAI or the reverse, and the cut-off GAI.

OptRx™

Aim
To determine changing nitrogen application requirements in-season and provide VRA prescriptions.

Technique
The system incorporates three optical measurement channels that can be used to construct two different vegetation indices – the normalised difference vegetation index (NDVI) and the NDRE.

OptRx uses the virtual reference strip concept. The healthiest area of the field is identified by the user and the area is scanned in real time. The VI calculated from this portion of crop represents plants with sufficient available nitrogen; it is called the VI reference value. This process locally calibrates the OptRx crop sensors to vary nitrogen application for the given growth stage. Ag Leader suggest that the OptRx be used for side-dressing corn at growth stages V5–V12 and wheat anytime between tillering and stem elongation. In this operational mode, the user can define minimum and maximum nitrogen rates to keep the system working within an input budget.

The system can also be used to vary nitrogen application to match a user-defined optimal nitrogen uptake amount for the crop growth cycle. The desired optimal nitrogen uptake is set, along with any known values for residual soil nitrogen and previously applied nitrogen fertiliser. By deducting the known soil and applied nitrogen amounts from the desired uptake, the potential nitrogen requirement can be calculated and compared with the calibrated VI measurements of the changing crop condition in the field. Nitrogen is then applied at rates to fulfil this mass-balance approach.

Greenseeker®

Aim
To determine changing nitrogen application requirements in-season and provide VRA prescriptions.

Technique
The system uses a process of comparing the reflectance (calculated as NDVI) from unfertilised plants to those in a non-limiting nitrogen strip to gain a response index (RI). The procedure involves the use of a series of generic functions in an algorithm which is made region- or site-specific by the inclusion of calibration coefficients obtained from local observations.

Essentially, the process involves:

- the prediction of potential yield without fertiliser (YP 0) from the NDVI and information on the growing degree days (GDD) or time since sowing;
- calculation of RI, which then is used to predict yield with added fertiliser (YP N);
- from the difference between YP 0 and YP N, the nitrogen fertiliser deficit can be calculated and a nitrogen rate determined.

N-Sensor® (ALS)

Aim
Primarily used to determine changing nitrogen application requirements in-season and provide VRA prescriptions, but also calibrated for the VRA of growth regulators.

Technique
The N-Sensor® (ALS) is a progression from the original N-sensor® that used ambient light as the source for reflectance measurements. The system calculates a biomass index using reflectance measured at the red edge of the spectrum, which is used to infer nitrogen uptake by a crop. Generally, the system can be used to determine changes in required nitrogen application rates in two ways:

- comparison of real-time reflectance to that from a previously measured nitrogen-rich strip or an optimal treatment section with a known nitrogen level;
- comparison of real-time reflectance to that from a number of initial field transects to variably apply a predetermined field average application.

The system has calibrations for a wide range of crops, but essentially it is designed to apply more nitrogen fertiliser to areas of the crop where the biomass index determines the nitrogen uptake of the crop is lowest. A biomass cut-off is used to identify areas where extremely low biomass is probably caused by conditions other than low soil nitrogen status. Fertiliser is reduced to avoid wastage in these areas. This level can be adjusted.

This system can also be used to apply growth regulator in cereals, to avoid crops lodging, based on the crop biomass relative to a predetermined maximum. Basically, the more biomass detected by the system, the more growth regulator will be applied to ensure a uniform concentration per leaf area. Areas with less biomass are not prone to lodging but may suffer yield losses due to overapplication and therefore the application rate will be reduced.

CropSpec™

Aim
To determine changing nitrogen application requirements in-season and provide VRA prescriptions.

Technique

The CropSpec™ was developed in conjunction with Yara International ASA, the developers of the N-Sensor®. The sensors mount in a similar fashion but the CropSpec™ system uses pulsing laser diodes as a light source and measures reflectance in slightly different spectral bands. It calculates a biomass index which is used to infer nitrogen uptake by a crop. The system can be used to determine changes in required nitrogen application rates in two ways:

- a simple two-point calibration is built by using the sensors to measure the lowest- and highest-performing areas in a field as determined by the user. The user then sets the application rate to apply for the field average and the sensors will vary application rates around that level based on observed variability in the index;
- subscription to optional software from Yara International ASA provides crop-specific calibrations to determine actual nitrogen requirements from the observed variation in the biomass index.

Like the other systems, the CropSpec™ can be used to record and map the observed reflectance for further analysis or provide real-time prescription data to drive on-the-go VRA of nitrogen.

Weedseeker®

Aim

The system is designed to minimise the amount of herbicide applied to fallow fields or, when used in conjunction with spray hoods, to row crops in-season.

Technique

The system measures the reflectance of red and infrared light and calculates NDVI. The system operates by recognising the difference between the NDVI of bare soil or dead plant residue and that of living plants. In essence, it identifies green objects against a brown/yellow/red background then uses that information to activate a series of boom-mounted chemical application nozzles.

WEEDit®

Aim

The system is designed to minimise the amount of herbicide applied to fallow fields.

Technique

The sensors use a red LED light source to induce fluorescence from the chlorophyll of green plants. This allows the system to identify living plants then use that information to activate a series of boom-mounted chemical application nozzles.

Cropping simulation

Cropping simulation uses computer-based models to predict cropping outcomes. There can be one or a number of outcomes predicted, depending on the purpose of the model. They are mostly used in SSCM to predict crop yield, but the more sophisticated simulators can provide predictions on crop, soil and water conditions over time.

The models themselves are mathematical formulas, or groups of mathematical formulas, that form the rules for combining data on the parameters required for the predictions. They can be derived simply from analysing the outcome of past crop experimental data (empirical) or they can be built to represent the processes that are involved in crop growth (mechanistic).

The accuracy and suitability of all models depends on how well they have been calibrated to real outcomes. The mechanistic simulation models are generally more complex and aim to predict a number of outcomes given a set of initial conditions, fertiliser and management inputs and weather scenarios. Each of these aspects is described by parameters in the models. The number of parameters, the degree to which they can be varied and the ways they interact depend on the sophistication of the model.

The benefits of mechanistic cropping simulation for SSCM are:

- they allow seasonal weather conditions to be incorporated into the prediction of outcomes;
- they predict outcomes for different scenarios by altering the value of initial soil, management and timing parameters. This provides the possibility of asking numerous 'what if?' questions;
- they provide predictions for various stages in the cropping process;
- they allow predictions using past conditions to build a library of outcomes over time.

The downside of mechanistic cropping simulation for SSCM:

- they require a lot of data to fill values for the parameters;
- the data requirements mean that they are used only on points in a field, not across whole fields;
- some component models may not be sensitive enough or well enough calibrated for all growing regions;
- they may not include a model for a process or condition that is important in some locations.

The most advanced and widely known crop simulation model used in Australia is the Agricultural Production Systems Simulator (APSIM). It comprises numerous models of the processes involved in establishing and growing a crop. These can be divided into:

- a set of biophysical models that simulate biological and physical processes in farming systems;
- a set of management models that allow the user to specify the intended crop management that will be applied to the simulation;
- a simulation engine that facilitates communication between the all the models and organises the simulation.

APSIM is quite detailed and time-consuming to learn, so it may not be suitable for most growers and advisers. However, the APSIM models have been incorporated into a more user-friendly web-based subscription service known as Yield Prophet® that is designed to assist management decisions during the cropping season. Growers enter inputs at any time during the season to generate reports of projected yield outcomes showing the impact on crop yield (based on type and variety) of sowing time, nitrogen fertiliser management and irrigation.

Yield Prophet® does not generate recommendations or advice. It combines the APSIM models with field-specific soil, crop and climate data supplied by the user to generate information about the likely outcomes of farming decisions.

APSIM and Yield Prophet® do not take into account weed competition, pest or disease pressure, agrochemical damage, farmer error or extreme events (e.g. extreme weather, flood, fire).

The output from a Yield Prophet® simulation includes information on the:

- probability of exceeding a grain or hay yield based on current nitrogen status or with nitrogen in non-limiting supply;
- predicted timing of a crop reaching target growth stages;
- background information on historical frost and heat shock likelihood;
- distribution of plant available water;
- changing availability of water to growing roots;
- changing water stress status of the crop;
- nitrogen budget, including crop and soil nitrogen;
- changing soil nitrogen content;
- changing nitrogen stress status of the crop;
- probability of future waterlogging;
- projected crop development, water and nitrogen use for the next 10 days;
- changing status of seasonal rainfall based on historical averages;
- Bureau of Meteorology rainfall forecast for the next three months;
- the potential impact of the Southern Oscillation Index (SOI) and the El Niño-Southern Oscillation Sequence System (ESS) on rainfall and yield predictions.

Cropping simulations using these and other software programs (see Chapter 4) can be powerful tools for predicting and understanding temporal variability in aspects of crop growth. The development of cropping simulation models that can use the detailed spatial information gathered by the sensors used in SSCM as input data would be a great step forward in creating powerful decision support systems that can incorporate spatial and temporal data in crop management decisions.

Incorporating SSCM into farm management: a summary

The decision to move into some form of vehicle navigation aid is generally recommended as the initial step for incorporating SSCM, but the use of SSCM does not necessarily require such systems. Small farm size and turnover may preclude such initial purchases but not the possibility of improving the management of variability.

To take PA further on-farm and utilise the navigation aid equipment more fully, the impact of variations in landscape and soil, previous management actions and climate on the variability in crop production can be investigated. SSCM exists to facilitate such investigation and help improve management decisions if required.

Importantly, the amount of variability in resources and crop yield on individual farms and fields is unique to each site. This is crucial, as it means that no single management prescription can be defined for SSCM. But it also provides the greatest impetus for exploring the use of SSCM, because the best information for optimally managing each farm or field will be found by local investigation.

Figure 7.3 outlines a general incorporation strategy that can be employed in conjunction with local agronomic understanding and advice at the key exploratory and decision stages. A more specific, tactical decision tree is provided in Figure 7.22, where the colour coding identifies which decisions in the tree are relevant to each step in the general strategy. It requires a positive or negative answer to decide on an individual incorporation path. It begins with the realisation that crop yield may not be uniform across a field/farm and provides a well-researched set of actions to cover scenarios that may occur, in most Australian regions, with the knowledge available in 2013.

An important point in this decision process is the ability to quantify yield variability because it drives the initial motivation and is critical in assessing the results of any changes in management. To provide a plan that is potentially useful to all farmers, objective quantification of production using crop yield monitors or calibrated biomass imagery is recommended.

Some farmers may wish to use their local knowledge of resource and yield variability to bypass some of the stages described here or to target specific local variability issues that are regarded as important and easily identifiable (e.g. variable lime application based on soil boundaries). As long as people are comfortable with the assessment of the size of the initial yield variability, its cause and their ability to assess the impact of management changes, then less-technological aids may be applied.

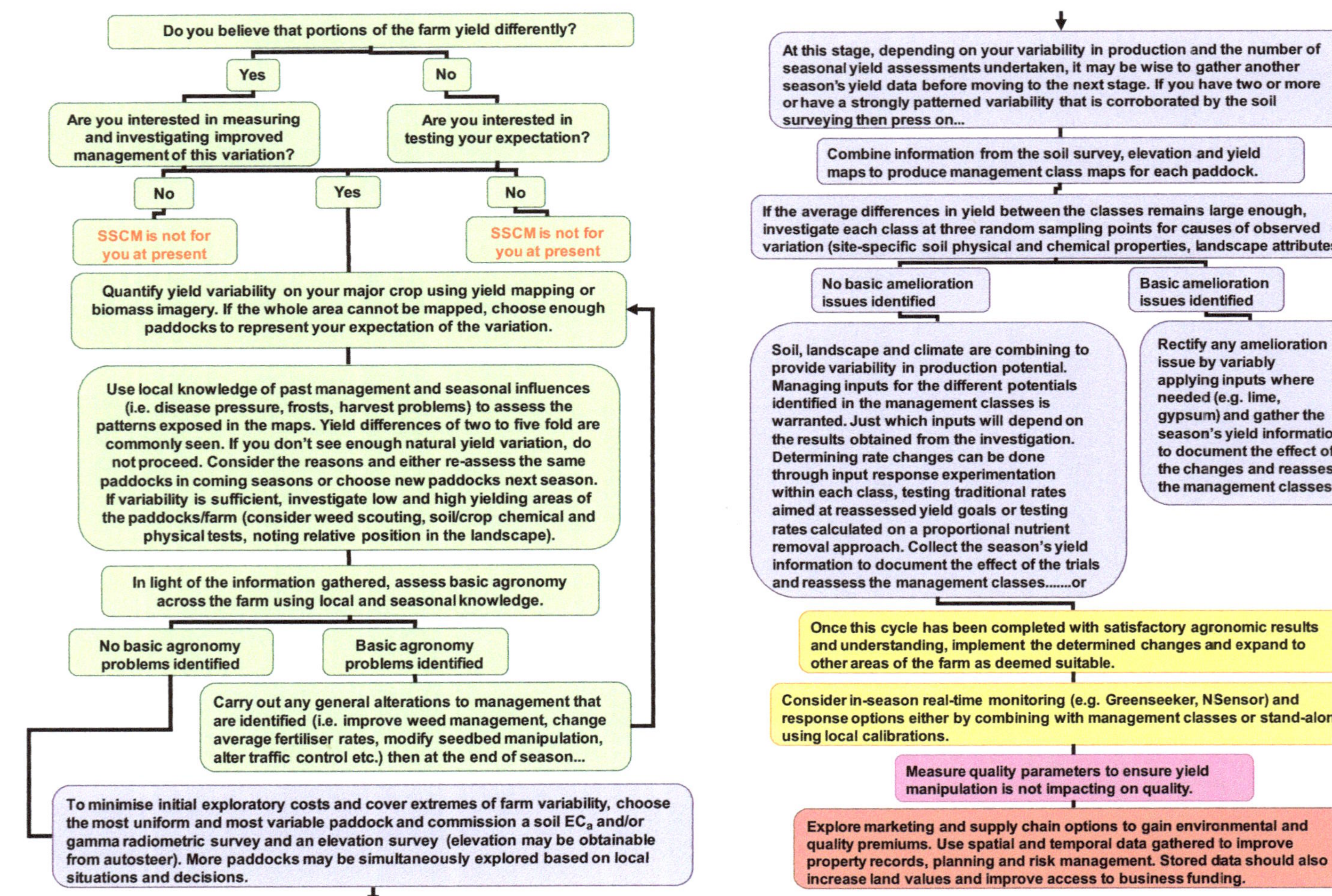

Figure 7.22: A tactical decision tree for incorporating SSCM in Australia.

However, for the most comprehensive benefits, it is recommended that physically documenting patterns and quantities of variation be undertaken for current assessment and decision-making, for auditing past actions and for future analysis. If this can be done at a resolution that is satisfactory to those involved without the use of some or all of the sensing systems identified as useful in this plan, then proceed without them. It is certainly possible to draw production class boundaries on maps and feasible to use manual switching or multi-pass application to achieve basic variable-rate applications if they are required. Employing variable-rate technology with positioning systems makes the job less stressful, however, and allows more sophisticated positioning and rate adjustments.

Key points

- Spatial variability in Precision Agriculture (PA) refers to the variation found in crop, soil and terrain properties across an area (e.g. field, farm).
- Temporal variability in PA refers to the variation in crop, soil and climatic properties within a certain area at different measurement times (e.g. different growth stages, successive seasons).
- In Australia, it is common knowledge that the most dominant influences on yield variability (other than climate and rainfall) are changes in soil physical factors such as soil texture, soil structure, soil depth and organic matter levels. These are known to contribute indirectly to the soil's moisture storage limits, cation exchange capacity and nutrient availability. It is important to explore variation in these factors when investigating the causes of crop yield variability.
- Site-specific crop management (SSCM) aims to better identify the changing yield potential within a field to improve decisions about the use of inputs such as fertilisers, agrochemicals, lime, gypsum and fuel to better match the changing requirements of the soil and crop. A better match should mean that inputs are used more efficiently, profits are maximised and waste is minimised.
- Before spending time and money exploring SSCM, ensure that the uniform-rate management options being used suit the average production potential of the farm. Correct any general issues with traffic and water management, soil pH or sodicity, fertiliser applications and weed and pest management.
- Variable-rate application (VRA) can be used to change the rate of inputs such as fertiliser, lime, seeds and agrochemicals.
- Variable-rate technology (VRT) is not necessary to achieve VRA. Using VRT just makes the job less stressful and allows more sophisticated positioning and rate adjustments.
- The potential benefits of VRA are generally higher when:
  - the amount of spatial variation is larger;
  - the pattern of spatial variability tends more towards coherent patches, usually meaning that fewer rate changes are required;
  - the pattern of variability is driven by spatial rather than temporal factors, so it is likely to be relatively stable from season to season and easier to formulate VRA plans;

– the unit cost of input is high relative to the price paid for the crop.

- A generic system of establishing potential management classes (PMCs) using high spatial density yield, reflectance, soil and terrain information has been shown to be useful across a wide range of agro-climates in Australia. Using this information to target sampling to understand the dominant causes of production variability and guide management decisions is becoming widely accepted.
- Creating more than four PMCs within a Australian grain field is highly unlikely to be economically viable, given present management techniques. Experience has shown that two or three different PMCs is the likely level of separation where class management is applied.
- The four main categories of VRA operations are:
  – map-based management class operations;
  – map-based whole-field operations;
  – real-time management class operations;
  – real-time whole-field operations.
- Map-based management class operations can be undertaken with a wide range of application equipment and applied to any input that is metered onto a field.
- Real-time, whole-field operations need specific hardware to be used for the input being controlled. At present these types of operation are widely used only to apply nitrogen fertiliser and herbicide. Some application of growth regulators is being undertaken.
- SSCM offers a step forward in the understanding and treatment of variability in field crop production. It is important to remember that the name says it all. Each field and each farm has a unique blend of landscape resources, climatic conditions, financial resources, past management effects and future management goals.
- Management decisions should not focus on treating a field to produce a uniform yield unless the potential is uniform. The benefits from SSCM analysis will only be realised by acknowledging diversity in yield potential and environmental conditions when formulating field management options. For example, well-documented areas of low yield potential may be removed from production, have their land-use changed or have their inputs reduced to minimise potential financial losses.
- Three general options to begin matching input rates to production variation are:
  – replacing nutrients removed;
  – modifying yield goals between areas or classes;
  – determining actual requirement differences between areas or classes.
- Replacing nutrients removed can be carried out on any crop but it is important to get an accurate estimate of the average quantity or concentration of the nutrient of interest in the crop being harvested. For nitrogen, the use of a protein monitor would make this process much more accurate.
- Employing an experimentation process will provide the most complete analysis of crop requirements, but requires a longer time-frame. It is worth pursuing because it gives an actual response that can be pinned to a seasonal climatic record and stored for the future. It can also be left in place and monitored over different seasons and crops to provide even more valuable information on the farm.

- As a reasonable trade-off between accuracy and intense data-gathering, it is useful to begin with site-specific information to modify yield goals between classes then to apply local agronomic estimates for input requirements across a farm. Experimentation in a number of fields could be implemented to hone goals and input estimates further.
- In the Australian dryland environment, it is not unexpected that factors controlling the interaction between crops and the climatic environment should be influential in the variability displayed in crop yield maps. For management, this suggests that in the future it will be beneficial to use information on management classes in conjunction with early season environmental indicators and crop response models (or simpler, empirical budget models) to guide variable-rate decisions.

8

Economics of PA in Australian grain crops

The economics of Precision Agriculture (PA) should be considered in a whole-farm context, just as all other aspects of farm investment. In PA, the analysis of investment outcomes is often confined to a financial balance sheet because it is simple. This approach certainly provides information to support decisions, but it doesn't encompass the broader notion of whole-farm economics. A full analysis would include the impacts on time and labour use requirements, as well as the environmental, job satisfaction and social outcomes. Such an analysis is difficult to perform using a single measurement scale such as money. Australian farmers should consider the balance sheet approach to PA as a useful tool to be included along with broader considerations when making decisions about the implementation of PA.

General introduction

Obviously, farming is not undertaken to intentionally lose money and in general this is not the case. But if farming operations are considered over only a short time-frame (e.g. a growing season), then financial losses can occur. Precision agriculture as a form of farm management will be no different, but the risk of short-term financial losses may be minimised by optimising the use of inputs to produce outputs. Profit may also be made from long-term improvements in operability, landscape and environmental management, product marketing,

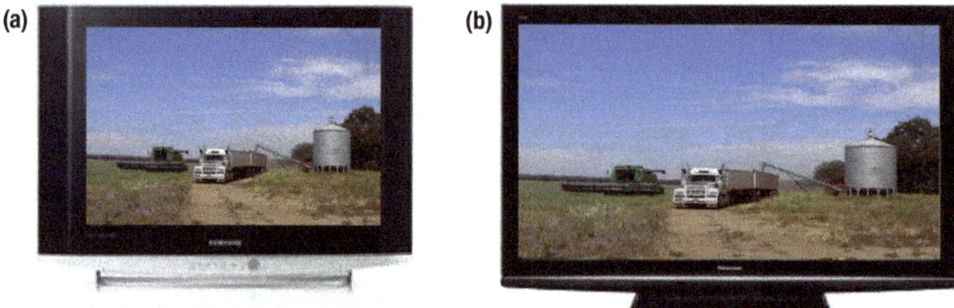

Figure 8.1: (a) A 30 inch CRT television. (b) A 50 inch plasma television. The plasma is more than three times the price of the CRT but both will function as a television. Deciding to invest in the plasma TV involves decisions on more than just function and finance.

storage of knowledge relevant to enterprise management and the farmers' contribution to improving society.

The economics of investment in agriculture is usually broken down into financial costs and returns. However, investment decisions are not always about basic function and finance. An individual investment decision can include the assessment of benefits that cannot be (easily) allocated a financial value.

Consider the purchase of a television. Figure 8.1 shows two different types, but the plasma TV is three times as expensive as the cathode ray tube model. There may be a difference in the quality of the job performed, ease of use and extended functionality, but both will display television broadcasts. The decision to invest in the more expensive television comes down to the desires, goals, financial status, risk profile and satisfaction assessment of the individual purchaser. The same issues can be linked with agricultural investments.

There are many different ways that PA can be incorporated into farm management, and it is unlikely that all potential PA applications will be used for any single production system. The diversity in production systems also means that the suite of PA tools that best suit one farm enterprise may be different from those that work on other farms. This makes the level of investment, and the return on investment, region-, farm- and even field-specific.

It is important to note that any economic analysis is influenced by contemporary capital/input costs and commodity prices. In agriculture especially, the specific relationships between costs and returns varies with time. As an example, the common decrease in the cost of a technology over time means the financial analysis of an early-adopting grower is not likely to reflect the situation two to three years in the future.

All financial analysis provided here should be used as a guide only. All prices are quoted in Australian dollars unless specified.

Balance sheet approach

The balance sheet approach provides a reliable method of incorporating the obvious financial costs and benefits into an assessment of PA investment outcomes. These obvious items are:

- purchase or contract price;
- amortisation and discount rate;
- opportunity cost;
- changes in applied inputs;
- changes in production income.

Spreadsheet tools to help build a balance sheet analysis are available from a number of sources and are generally applicable to the investment in GNSS-based vehicle navigation and VRT. Adding up the major financial costs for an investment provides a good idea of the size of the financial investment and allows for comparisons between different operational combinations. When these costs are put into a 'dollar per hectare' basis ($/ha), it is easy to get a feel for the costs in comparison to other farm-wide investments (e.g. tractor ownership and operating costs).

Knowing the costs also provides an idea of the 'benefit' that is required to break even in financial terms. The benefit is often equated to a required reduction in input use or increase in crop yield in $/ha. However, the main issue here is how to gauge the extent of potential benefits in these two areas, when the variability of each farm and its management will be the determining factor. On top of this is the prospect of achieving benefits in other areas that may be difficult to put into $/ha terms.

General information about costs of PA

Costs will depend on the degree of adoption, equipment purchased and amount of support required. Table 8.1 provides a general guide to the major costs in Australian dollars (2012 prices). Growers should contact local equipment and service providers regarding the specific costs for their production system.

Several studies have looked at the specific cost of adopting PA. The figures from Australian farms and adjusted values from the USA generally indicate that the cost will be A$5–25/ha for a simple PA system, rising up to A$60/ha for a more advanced system on a smaller farm. These figures are only a guide; costs will depend on which aspects are adopted and to what degree. Growers are advised to compile a cost guide for themselves based on determined requirements and local costs.

Annual fees

Some PA technology comes with an annual licensing fee. Such fees are most commonly applied to correction signals for GNSS and data management software.

Table 8.1: General range of total and amortised costs for the major PA investments

PA-related item	Cost range (A$)	Potential use (years)	Costs for 1000 ha annual cropping area (A$/ha/year)
Vehicle navigation			
Light bar	3500–5000	7	0.60–0.86
Autosteer 10–30 cm	15 000–25 000	7	2.57–4.28
Autosteer 2 cm, own base station	35 000–45 000	7	6.00–7.71
In-field monitoring			
Yield monitor	4000–6000	7	0.69–1.03
Protein monitor	15 000–25 000	7	2.57–4.28
Proximal crop/weed reflectance sensors (4 units)	8000–12 000	7	1.37–2.06
Data management software	2000–5000	5	0.34–0.86
External monitoring and consultants			
Remotely sensed crop reflectance imagery	0.50–10/ha	1	500–10 000
Soil EC_a and/or gamma-radiometric survey	3–5/ha	25	0.12–0.20
Soil and crop sampling and analysis	50–200/sample	3	Depends on sampling scale
Consultant/analyst	3–15/ha	1	3000–15 000
Variable-rate application			
Variable-rate components for air-seeder/spreader	5000–30 000	7	0.86–5.14
Variable-rate components for chemical application	5000–15 000	7	0.86–2.57

While these may seem small on an annual basis compared to some initial upfront costs, the fees will accrue over time. They should be budgeted into any purchase.

General information about return on investment for PA

In a typical cropping enterprise, inputs such as fertiliser, chemicals, seed and labour make up two-thirds of the variable costs. Using PA to reduce some of these costs is the simplest way to gain a return from a PA investment. Using these inputs more efficiently to produce a higher input to yield ratio, increases returns further.

In reality, the financial benefits of PA are likely to come from a mix of savings and improved efficiency.

Vehicle navigation aids (autosteer and guidance)

Vehicle navigation aids (using guidance, steering assist or autosteer) allow growers to reduce areas of application overlap or misapplication during field operations. This produces a benefit through decreased input application and increased operational efficiency.

Application overlap using conventional marking tools can be anywhere from 0.5 m to 1.0 m (i.e. 2–4% on a 27 m wide implement). Reducing overlap to 2 cm (0.07% on a 27 m wide implement) essentially produces savings in input costs (fuel, fertiliser, agrochemicals etc.) of 2–4%. On narrower implements, the savings are larger as the overlap will be a larger percentage of the implement width and more passes are required in a field. Therefore, the size of the benefit on each farm will depend on the quality of vehicle navigation prior to adoption, the operational specifications of farm machinery and the navigation technology purchased.

The impact of these savings on the farming gross margin will depend on the contribution of each input to variable costs on each farm, but generally the improvement in gross margin means that the cost of any investment in autosteer/guidance is recouped over a few seasons. On top of this, there are other agronomic benefits from adopting high-precision autosteer systems. These include:

- improved soil condition away from wheel tracks;
- inter-row sowing options;
- increased opportunity for operational timeliness.

The value of these will vary from field to field and farm to farm depending on the production systems. For example, in the Wimmera region of Victoria, yield increases of 10–20% have been reported in lentils sown directly in the inter-row of standing cereal stubble. The lentils are able to use the cereal straw for support, thereby reducing lodging and increasing harvest efficiency. Similarly, direct sowing in the inter-row has been shown to reduce the effect of soil-borne diseases on yield in second-year wheat–wheat rotations. Autosteer/guidance permits operations to be performed at night or in poor visibility (e.g. fog), which may lead to more timely operations. Being able to ensure operations such as sowing and chemical application are performed at the optimum time can dramatically affect yield potential.

Variable-rate spray application

Spray costs can be reduced by applying variable rates of chemical across a field. Identifying the locations of weeds and focusing treatment in those areas can dramatically reduce the amount of chemical applied to a field, compared to a blanket application. It is rare that more than 40% of a cropping field is infested with weeds; at least 60% of most fields require zero or reduced treatment.

The location of the weeds can be mapped prior to treatment or identified during the treatment operation using reflectance sensors. In either case, it is the density of the weeds and their location in each field that controls the size of the potential benefit from variable-rate herbicide application. As an example, Tables 8.2 and 8.3 show savings made on commercial cropping operations using the WeedSeeker® reflectance system in spray operations. Table 8.2 shows figures for individual fields where herbicide application was required on only 4.5% to 15% of the field area. Table 8.3 records the outcome of spray operations over a number of seasons, across a whole farm. On that large operation, the cost of the spray program was more than halved during the four seasons.

Table 8.2: Operational savings using the Weedseeker® to spot-spray weeds in single fields

	Field 1	Field 2
Seasons	1	1
Area blanket sprayed (ha)	246	120
Weeds	Peachvine, milkthistle, fleabane, volunteer cotton	Volunteer cotton
Herbicide	2.6 L/ha Rup + 4 L/ha Surpass	1 L/ha Starane + 1 L/ha MCPA
Area sprayed (ha)	11.8 (4.5% of total)	18 (15% of total)
Cost of blanket-sprayed herbicide (A$)	7840	3360
Cost of spot-sprayed herbicide (A$)	353	504
Average saving (A$/ha)	30.43	23.80
Total saving (A$)	7487	2856

Data courtesy of David Brownhill, Merrilong Pastoral Company.

Identifying variation in financial return

Traditionally, gross margin analysis, which reconciles income against variable costs, has been used to assess the financial return from cropping operations on a field-by-field basis. With the collection of crop yield maps, the process can be taken to the within-field level by multiplying the spatial yield data by the crop price. This produces spatial income data which, combined with the variable costs for a field, results in site-specific gross margin maps.

In Australia, wheat and barley sales attract premiums or discounts based on protein classes and delivered grain moisture content. By adding protein and moisture map data to the site-specific gross margin mapping process, a comprehensive spatial map of gross margin that incorporates the seasonal premium/discount payment matrix can be built for a field. Figure 8.2 shows the maps used, and produced, in applying this process to a 40 ha field of wheat.

Table 8.3: Operational savings using the Weedseeker® to spot-spray a variety of weeds over four seasons

	Whole farm
Seasons	4
Total area to be sprayed (ha)	27 388
Average area sprayed (% of ha)	17
Average rate (L/ha)	1.5
Average cost (A$/ha)	5
Average saving (A$/ha)	6.5
Total saving (A$)	178 022

Data courtesy of David Brownhill, Merrilong Pastoral Company.

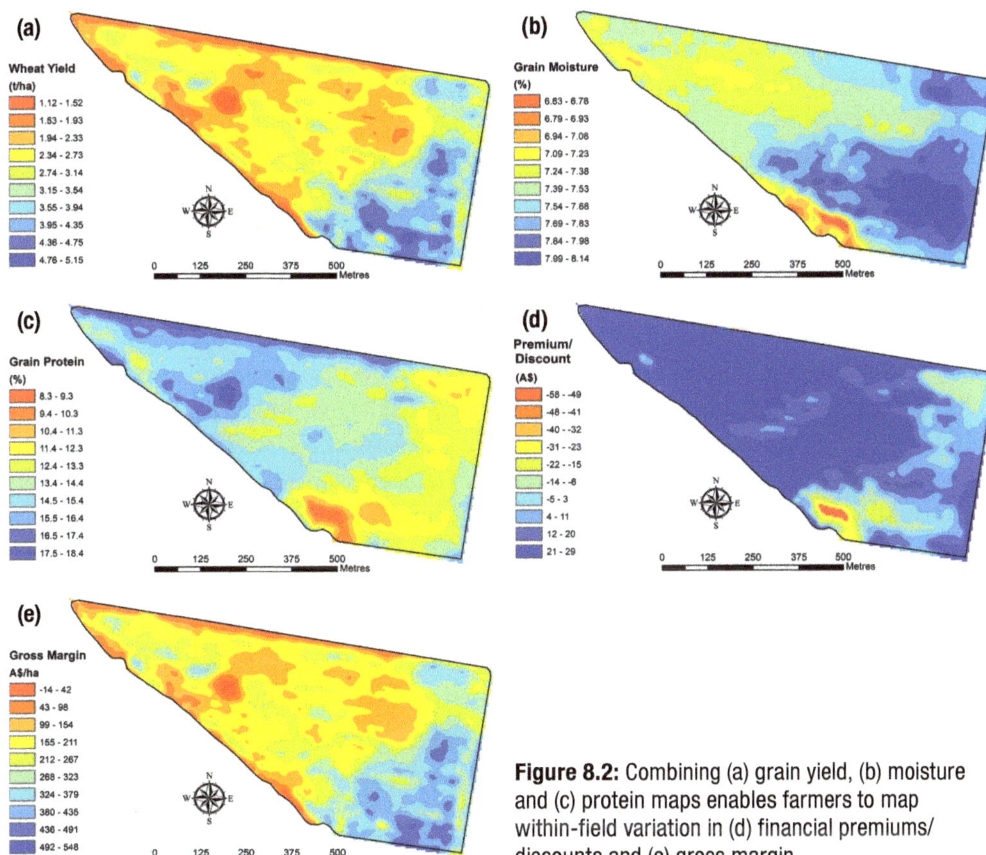

Figure 8.2: Combining (a) grain yield, (b) moisture and (c) protein maps enables farmers to map within-field variation in (d) financial premiums/ discounts and (e) gross margin.

Substantial variation in gross margin within the field is certainly evident. While crop yield remains the major driver of revenue in this field, being able to map the gross margin using the major parameters of the payment matrix provides unique financial information that can inform a range of strategic and tactical management decisions for the field.

Managing yield variation

When presented with variation in yield, the first inclination is often to try to make the production more even, notably by trying to 'improve' the lower-performing areas of the field. PA tools and techniques can be used to identify low-performing areas and help pinpoint the causes of poor yield and where they can be rectified by management operations; in such cases, improvement may be possible. The application of soil ameliorants such as lime and gypsum fits this scenario.

However, areas of low production caused by natural properties of the soil or location cannot be altered in a cost-effective manner. Soil type, texture and position in the landscape fit this scenario. Here, identifying the yield potential of each area is important and management options should be devised to optimise production at each location. Often it is found that the level of inputs traditionally applied in blanket-rate applications across a whole field suffice for maximum production in low-production areas, and savings may be made by reducing applications. At the same time, increased returns can often be found by identifying and optimising inputs into the best-performing areas of a field. These areas may not have reached their full production potential under blanket-rate field management.

Variable-rate ameliorant application

Lime and gypsum are used to correct soil pH or structural problems, respectively. The application rates are usually based on a representative soil analysis result for a whole field. The actual requirement for these ameliorants is usually closely linked to soil type; where soil type changes within a field, the optimum application rate will usually change. Both products need to be applied at the correct rates to be effective at improving yield potential.

Identifying the correct rates for different areas within a field using strategic sampling may reduce the amount of ameliorant, compared to a blanket-rate application. Where more ameliorant is required than traditionally applied, the cost will be outweighed by the improved yield potential. Specific financial benefits will depend on the variability in soil type and previous management practices on every farm.

As an example, Figure 8.3 shows an EC_a map and the subsequent classification of the field into two potential management classes (PMCs). Sampling for soil pH within those classes produced the results in Table 8.4, showing that 83% of the field required no lime application. If the field had not been sampled by PMC and soil from only Class 1 had been used to represent the whole field, that saving would have been lost. If soil from only Class 2 was used to represent the field, then no lime would have been applied and a potentially significant yield loss in Class 1 may have occurred in the current and future seasons.

Variable-rate fertiliser application

The potential financial benefits associated with variable-rate fertiliser application are also specific to each farm and field. The value of variable-rate fertiliser

Table 8.4: Soil pH results, lime recommendations and treatment costs for the field in Figure 8.3

	Size (ha)	Topsoil pH	Lime recommended (t/ha)	Cost @ A$50/t spread (A$/area)	Cost of whole field treatment at pH 4.8
Class 1	16	4.8	1.3	1040	1040
Class 2	79	5.7	Monitor	0	3710
Total	95			1040	4750

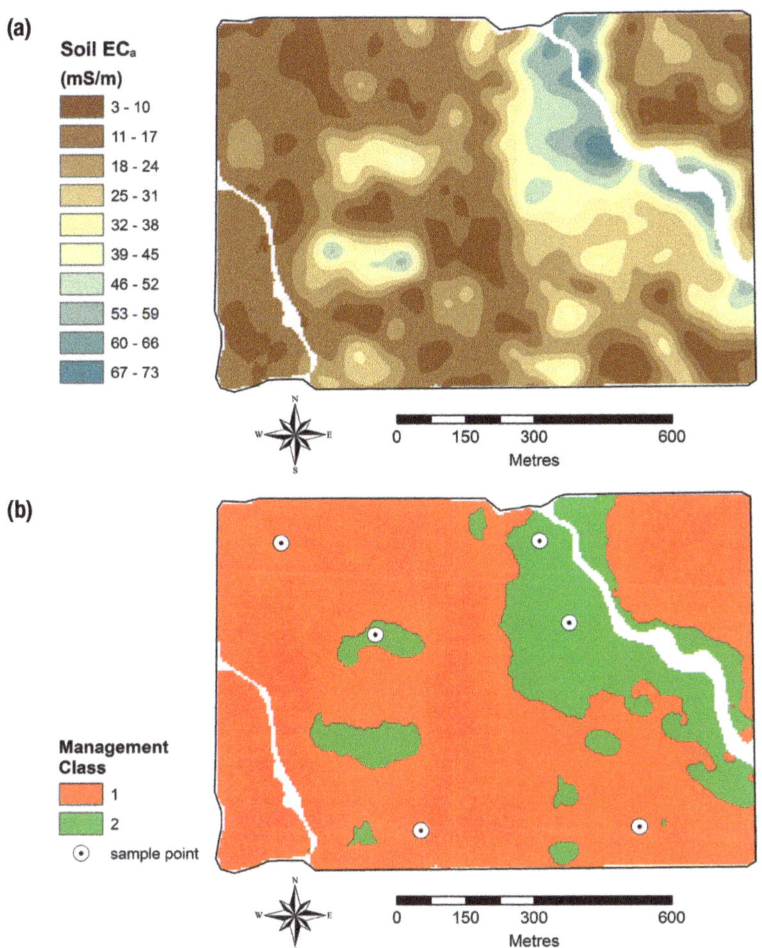

Figure 8.3: (a) Soil EC_a map and (b) PMC map used for a soil pH sampling operation.

application will depend on the amount of variation in soil type, landscape, inherent nutrient reserves within a field and prices for fertiliser and crop product. The suitability of the previous uniform-rate applications and fertiliser timing strategy (e.g. all up-front, split applications, whether up-front and split applications are variable-rate) will also affect the size of potential financial benefit.

The best way to measure the size of a potential benefit is to conduct response trials to determine whether there are different optimum nutrient application rates for areas within a field. Armed with management class yield response information from in-field experiments, it is possible to compare the gross margin under standard management to the gross margin achievable under optimum-rate management. The gross margin for each scenario is calculated by multiplying the yield by the grain price and deducting the cost of fertiliser. Any difference between the two gross margin figures is termed a 'net wastage' of standard management,

Table 8.5: Gross margin analysis comparisons between standard uniform-rate application of phosphorus fertiliser and optimum-rate management

Year	Size (ha)	Crop	Yield (t/ha)	Net wastage (A$/ha)	Proportion of season P fertiliser costs (%)
2003	40	Wheat	4.4	55	93
2003	110	Wheat	2.2	36	129
2004	34	Wheat	1.8	50	85
2004	40	Faba beans	2.0	50	85
2004	110	Field peas	1.0	8	30
2005	34	Barley	4.5	39	74
2005	39	Wheat	4.3	61	105
2005	40	Wheat	5.6	24	43
2005	110	Wheat	3.1	65	236
2006	55	Wheat	0.9	36	78
2006	110	Barley	1.0	33	121
2007	39	Wheat	1.5	103	177
2007	43	Canola	0.9	58	169
2007	55	Canola	0.5	18	53
2007	91	Wheat	1.1	45	154
2008	39	Wheat	1.4	59	140
2008	43	Wheat	2.3	77	189
2008	55	Wheat	1.2	26	40
Minimum	34			8	30
Median	43			48	99
Maximum	110			103	236

because in essence it is dominated by direct costs associated with any overapplied fertiliser, any yield losses due to less than optimal fertiliser application and, depending on the shape of the response functions, yield losses due to any overapplication of fertiliser.

Table 8.5 shows the results of this type of analysis for 18 yield responses to phosphorus fertiliser experiments over a number of seasons, crops and sites across Australia. Table 8.6 shows the results for 15 yield responses to nitrogen experiments. The tables show that there was always a net wastage in gross margin from standard management in all the fields, with a median wastage figure of A$48/ha for phosphorus fertiliser and $39/ha for nitrogen fertiliser.

These figures are a function of fertiliser costs and crop prices obtained in the specific year of each trial. They are also obtained from production systems in a variety of different regions, with different yield potentials, fertiliser application rates and seasonal climatic conditions. So the net wastage calculated, e.g. $10/ha,

Table 8.6: Gross margin analysis comparisons between standard uniform-rate application of nitrogen fertiliser and optimum-rate management

Year	Size (ha)	Crop	Yield (t/ha)	Net wastage (A$/ha)	Proportion of season N fertiliser costs (%)
2003	47	Wheat	4.8	4	12
2003	50	Wheat	3.2	48	372
2003	130	Canola	2.3	12	30
2004	22	Canola	1.7	79	143
2004	43	Wheat	3.3	46	130
2004	50	Barley	2.4	7	53
2004	79	Wheat	4.5	25	45
2004	80	Wheat	4.9	15	26
2004	130	Wheat	2.5	39	97
2005	43	Barley	2.5	28	80
2005	130	Barley	4.4	74	177
2006	50	Wheat	2.0	51	334
2006	130	Canola	0.4	20	43
2008	97	Wheat	3.5	103	240
2008	130	Barley	2.5	78	115
Minimum	22			4	12
Median	79			39	97
Maximum	130			103	372

will have a different inference for different production systems. In an attempt to overcome this, Equation 7.5 (see Ch. 7) was used to standardise the net wastage in A$/ha for each field to the A$/ha cost of nitrogen or phosphorus fertiliser actually applied across the field using the traditional uniform rate. In this way, the information derived from within-field trials can be more easily compared across all the production systems and seasons.

The proportion of seasonal fertiliser cost is therefore an estimate of the wastage as a proportion of the traditional investment in the specific fertiliser each year. The median value of 99% for phosphorus and 97% for nitrogen suggests that the potential financial benefit over a number of seasons gained by knowing more about the optimum rates of fertiliser for a field could be equal to the amount of money traditionally outlaid on fertiliser. For example, if the average phosphorus inputs were A$40/ha/year, then maximum improvements in gross margin of the same amount per year may be possible.

Obviously, the seasons affect the yield results and the gross margins. The response functions for many of the PMCs across the trials were relatively flat during the 2006–08 seasons, when annual rainfall was generally in the lowest

10–20% of years recorded and the in-season rain remained well below average in most of the areas. These types of seasons would be expected to provide poor return to fertiliser outlay, reflected in the low yields and high proportions of fertiliser losses reported for those years in Tables 8.5 and 8.6. However, even in 2005 when annual and in-season rainfall was generally above average, significant wastage in fertiliser application was still documented.

It is important to understand that the optimum fertiliser rates were determined at the end of the season. Choosing the optimum rates at the time of application is difficult; Chapter 7 discussed options for using PA data to help make those decisions. However, the figures discussed here show that there are worthwhile financial gains from improving the targeting of fertiliser application rates to crop requirements.

Other benefits

It is worth considering the impact of potential benefits that are difficult to show in a balance sheet approach. The values will be specific to each farming operation and may not be calculable in dollar terms. Benefits may include:

- increased speed of operations;
- improved timeliness of operations;
- improved ease and efficiency of operations;
- working more hours/shifts safely;
- greater flexibility in use of labour;
- potential quality increase;
- options for commodity differentiation on quality;
- options for commodity tracking/preservation of identity;
- potentially reduced chemical storage and handling;
- spatial recording of operations for future management use;
- spatial recording of operations to avoid litigation;
- spatial recording of operations for insurance claims;
- increased farm enterprise value with spatial records;
- reduced erosion potential;
- reduced environmental footprint;
- identifying areas for land-use change;
- facilitating carbon auditing based on production variability;
- increased peace of mind/management confidence.

Triple bottom line accounting

Triple bottom line accounting involves expanding the traditional accounting framework to take account of environmental and social performance in addition to financial performance. While it is often difficult to put a dollar value on environmental and social benefits, it is the environmental benefits of SSCM that are likely to be the next contributors to economic benefits in Australia.

There are two main concepts in the potential environmental benefits of SSCM that may be transformed into economic benefits:

- reduced pollution – minimising the escape of nutrients and chemicals from the farm system into the environment. Fines or regulations limiting the use of these products in certain areas would assign an economic value to the efficient use of these resources, while optimising production. This already occurs for the management of nitrogen in a number of European countries. Threshold values for water and air quality are used to provide limits that must be met. The challenge is to minimise the costs of achieving those limits by the creative use of innovative management techniques, including SSCM;
- product differentiation – products with enhanced attributes, such as perceived better quality (due to fewer chemicals used), environmental friendliness (reduced environmental impact) and traceability would be expected to command a price premium over a conventionally produced product. The use of SSCM offers the ability to fulfil such requirements.

It is highly likely that environmentally (and socially) responsible growers will be rewarded in the future for good management practices. Currently, authorities focus on penalising degrading practices; there is poor understanding of the monetary value of benefits, for both the grower and society, from sustainable practices. There is movement towards the concept of payments for ecological services provided by farmers (e.g. the recent federal Carbon Farming Initiative provides some options for environmental payments), but in Australia ecological payments generally remain outside the accounting process.

Whole-farm assessment

A number of surveys have been undertaken in Australia in an effort to quantify the financial outcome of PA investment on a whole-farm scale. It is a difficult task due to the necessary assumptions about the extent of application across the farm, the real source and size of contributions to changes in financial returns and the effective life of the equipment over which to spread the cost. It is also difficult to transfer the conclusions because, as with the field-scale analyses described previously, the outcomes are based on costs and prices at the time of investment or sale.

However, two surveys that have provided worthwhile attempts at quantifying and partitioning whole-farm investment costs and benefits of PA across a number of individual farms are summarised here. The benefit figures they arrived at are similar, and within the range shown in the individual field experiments reported earlier in this chapter.

Economic benefits of PA: case studies from Australian grain farms (CSIRO)

This survey, undertaken by the CSIRO, involved six early adopters of PA on farms spread across the eastern and western grain-growing regions of Australia. At the field level the gross margin per field ranged from –A$28 to +A$57/ha/year. At the whole-farm scale, the benefits ranged from A$14 to A$30/ha, with an average of A$22/ha. At the whole-farm scale, variable-rate application (VRA) produced higher returns per hectare than improved vehicle navigation (overlap). In general, VRA (predominantly fertiliser) provided returns of A$12–22/ha across the production systems compared to A$1–8/ha for improved vehicle navigation (Table 8.7).

The farmers noted a number of benefits other than improvements from reduced overlap and VRA. These included more efficient harvesting, reduced fuel use, reduced soil compaction, more timely sowing, the ability to conduct trials, increased knowledge of field variability, increased confidence to vary fertiliser rates, and improved weed control. The authors concluded that the survey results 'demonstrate that Australian grain growers have adopted systems that are profitable and are recovering the initial capital outlay within a few years, and they also see a number of intangible benefits from the use of the technology'. The survey illustrated that the use of, and benefits from, PA technology varies from farm to farm, in line with farmer preferences and circumstances.

Farmer case studies on the economics of PA technologies (SPAA Precision Agriculture Australia)

This survey, conducted by SPAA Precision Agriculture Australia, comprised six early adopters of PA on farms across South Australia. At the whole-farm scale, PA technology was found to return benefits ranging from A$11 to A$37/ha, with an average of A$18/ha annually (Table 8.8). Improved vehicle navigation provided an average benefit of A$5/ha and the benefits of VRA averaged A$8/ha. Half the respondents noted additional financial benefits, attributed to inter-row sowing and reduced soil compaction.

The survey noted that all farmers valued improved vehicle navigation because it reduced fatigue, and one-third believed the ability to run and monitor on-farm trials improved decision-making.

Importantly, it concluded that, as the cost of PA technology continues to fall from the prices paid by the early adopters, the time to payback should decrease and profitability of adopting PA technologies should rise for those newly adopting.

Summary

This discussion and the data presented do not guarantee financial benefit. Together they show the scale of financial gain that may be achieved, against which the financial costs must be balanced, to determine a financial outcome. What this information does show is that such financial analysis must be treated site-

Table 8.7: Summary of the estimated investment, benefits and cost recovery period for six farmer case studies.

Location	Size (ha)	Crops	Outlay (A$)	Technology	Benefits (A$/ha)				
					Total	VRA	Overlap	Other	Years to payback
Casuarinas WA	2600	Wheat, lupin, barley	90 000	Guidance, VRA fertiliser	21	16	5		4
Cunderdin WA	5800	Wheat, lupin, barley	189 000	Guidance, autosteer, controlled traffic, shielded spray, VRA fertiliser	22	13	7	2	2
Buntine WA	3400	Wheat, lupin, barley, canola	65 000	Guidance, autosteer, controlled traffic, VRA fertiliser	21	12	1	8	2
Moree NSW	1250	Wheat, barley, sorghum, chickpea, canola, sunflower	55 000	Guidance, autosteer, controlled traffic, VRA fertiliser and pesticide	30	22	8		2
Gunnedah NSW	3430	Wheat, barley, faba, canola, sorghum, maize, sunflower	95 000	Guidance, autosteer, controlled traffic, VRA fertiliser, in-season reflectance	24	20		4	3
Barmedman NSW	4000	Wheat, canola	51 000	Guidance, VRA fertiliser, in-season reflectance	14		7	7	5
Average					22	17	6	5	3

From M Robertson, P Carberry and L Brennan (2009) The economic benefits of precision agriculture: case studies from Australian farms. *Australian Journal of Agricultural Research* 60, 799–807.

specifically because it will need field- or farm-specific data for an analysis of the true financial position. Understanding what may be gained on each farm then allows a more informed consideration of which of the many options (extra costs) of PA may be most suitable in each case.

However, the goal of PA is management progress on a whole-farm basis, and this should be the realistic point of financial assessment of PA.

Table 8.8: Summary of the estimated investment, benefits and cost recovery period for six farmer case studies

South Australian region	Annual rainfall (mm)	Area cropped (ha)	Capital invested in PA		Annual benefit		Years to payback	
			Total A$	A$/ha	Total A$	A$/ha	Navigation aids	VRT
Mallee	250	3000	68 500	23	32 850	11	4–5	1
Upper Eyre Peninsula	300	4475	52 000	12	47 842	10	5	na
Mid North	400	1600	98 500	62	20 180	13	1–5	10
Mid North	400	2340	34 432	15	35 100	15	1	6
Lower Eyre Peninsula	425	2700	73 000	27	57 240	21	1–2	na
Lower North	475	1200	73 800	62	44 880	37	3	9
Average	375	2550	66 705	34	39 682	18	3	7

From M McCallum (2008) *Farmer Case Studies on the Economics of PA Technologies*. SPAA Precision Agriculture Australia.

Key points

- In a typical breakdown of cropping gross margin analysis, fertilisers, agrochemicals, seed and labour make up around two-thirds of the variable costs. Using PA to reduce some of these costs is the simplest way to maximise the return from a PA investment.
- There are many PA operations that can lead to economic benefits. The major benefits come from two main areas: reduced use of inputs through improved vehicle navigation using a global navigation satellite system (GNSS), and identifying and managing changes in yield potential through site-specific crop management (SSCM).
- The acquisition of a high-accuracy GNSS is the most expensive single component of a PA system. A GNSS is used to increase the precision of vehicle trafficking and input application. This has the immediate economic advantages of reduced areas of application overlap or misapplication during field operations, and reduced fuel consumption.
- The equipment required to achieve basic SSCM (yield monitor, variable-rate technology (VRT) and software) are less expensive but require more time and consideration in their use.
- Benefits from PA adoption will vary from one farm to another. Any benefits from SSCM and the associated costs are, by nature, farm-specific and field-specific.
- PA technology has generally decreased in cost since its introduction and, if this trend continues, adoption is likely to increase.
- In addition to economic benefits, improved accuracy of input delivery has certain environmental benefits, such as the reduction of greenhouse gas emissions and offsite contamination from agrochemicals or fertilisers. Automated navigation also

provides social benefits by reducing driver fatigue, stress and error. It allows the operator to concentrate on the application equipment (seeder, sprayer etc.) rather than on the vehicle.

- To adopt and use PA technologies effectively, producers have to make financial investments in learning skills as well as technology.
- For most growers, the benefits of PA are likely to come from a mix of savings in spraying costs, fertiliser costs and machinery costs and from the increased grain yield.
- Environmental and social benefits are difficult to put a dollar value on, but will become more significant as PA evolves in Australia.

Further reading

The knowledgeable application of PA involves the study of topics across a wide range of disciplines. This short list includes books, a manual, a scientific journal, proceedings from two continuing PA conference series and two volumes of practical case studies, that readers can use to gain an increased depth of understanding in subjects that are important to PA.

El-Rabanny A (2006) *Introduction to GPS: The Global Positioning System*. 2nd edn. Artech House, Boston.

Grains Research & Development Corporation (2006) Precision Agriculture Manual. GRDC, Canberra. <www.grdc.com.au>.

Isaaks EH and Srivastava RM (1990) *An Introduction to Applied Geostatistics*. Oxford University Press, New York.

Jensen JR (2007) *Remote Sensing of the Environment: An Earth Resource Perspective*. 2nd edn. Prentice Hall, New Jersey.

Marschner H (1995) *Mineral Nutrition of Higher Plants*. 2nd edn. Academic Press, London.

Mayfied A, Murphy K, Bramley R and Sargent M (Eds) (2008) *PA in Practice*. SPAA Precision Agriculture Australia, Mildura.

Nicholls C and McMallum M (Eds) (2012) *PA in Practice II*. SPAA Precision Agriculture Australia, Mildura.

Oliver MA (Ed.) (2010) *Geostatistical Applications for Precision Agriculture*. Springer, New York.

Peverill KI, Sparrow LA and Reuter DJ (Eds) (1999) *Soil Analysis: An Interpretation Manual*. CSIRO Publishing, Melbourne.

Precision Agriculture (1999–) Scientific journal published in six issues annually. Springer. ISSN 1385-2256.

Proceedings of the European Conference on Precision Agriculture (1997–) Published biennially by various entities from 1997.

Proceedings of the International Conference on Precision Agriculture (1992–) Published biennially by various entities from 1992.

Srinivasan A (Ed.) (2006) *Handbook of Precision Agriculture: Principles and Applications*. Haworth Press, New York.

Index

www.ingramcontent.com/pod-product-compliance
Lightning Source LLC
LaVergne TN
LVHW060630110826
845147LV00014B/885

* 9 7 8 0 6 4 3 1 0 7 4 7 2 *